MODERN STEAM PLANT PRACTICE

The Institution of Mechanical Engineers

MODERN STEAM PLANT PRACTICE

A Convention arranged by the
Steam Plant Group
of the Institution of Mechanical Engineers
and the Koninklijk Instituut van Ingenieurs
and held at 's Gravenhage
28–30th April 1971

1 BIRDCAGE WALK · WESTMINSTER · LONDON S.W.1

© The Institution of Mechanical Engineers 1971

ISBN 0 85298 048 5

ORIGIN	BKSL	O. N.	CST	L & D
CGW	S	LE 3109 29/11/72	5.40	ML/ML
CLASS No. 621.17/06		12FEB1973		K No.
AUTHOR				
SR		No.		SHL

D
621.18
MOD

CONTENTS

Modern Steam Plant Practice

A CONVENTION on Modern Steam Plant Practice, arranged by the Steam Plant Group of the Institution of Mechanical Engineers and the Koninklijk Instituut van Ingenieurs, was held at the Netherlands Congress Centre, Churchillplein 10, 's Gravenhage, on 28–30th April 1971. 124 delegates registered to attend. The Convention was formally opened by Ir P. Tuininga, Chairman of the Mechanical Engineering Division of the Koninklijk Instituut van Ingenieurs. The Opening Address was given by Professor Ir J. J. Broeze of the Technische Hogeschool, Delft.

The Convention was divided into three sessions at which papers were presented and discussed as follows:

Session 1. Chairmen: Mr A. Sherry and Ir R. Van Erpers Royaards. Papers C70/71, C71/71, C73/71, C75/71, and C97/71.

Session 2. Chairman: Professor Ir J. A. de Jong. Papers C64/71, C72/71, and C80/71.

Session 3. Chairmen: Mr A. W. C. Hirst and Professor R. W. Stuart Mitchell. Papers C61/71, C76/71, C84/71, and C89/71.

On 28th April there was a Reception at the Headquarters of the Koninklijk Instituut van Ingenieurs in 's Gravenhage. On 29th April the Convention Dinner was held in the Blauwe Zaal, Hotel Kurhaus, Scheveningen.

Technical visits were made on 29th April to the Rotterdam Dockyard Company and the Central Galileistraat GEB of Rotterdam.

The members of the Organizing Committee were: Mr H. Morris, C.Eng., F.I.Mech.E. (Chairman), Mr R. L. J. Hayden, B.Sc.(Eng.), C.Eng., F.I.Mech.E., Mr A. W. C. Hirst, B.Sc.(Eng.), C.Eng., F.I.Mech.E., Mr L. A. Robey, C.Eng., M.I.Mech.E., Mr A. Sherry, M.Sc., C.Eng., M.I.Mech.E., Mr B. Wood, M.A., C.Eng., F.I.Mech.E., and Ir P. Tuininga of the Koninklijk Instituut van Ingenieurs.

C61/71 AIR-POLLUTION PROBLEMS FROM LARGE POWER STATIONS

G. NONHEBEL*

The paper discusses the reasons why meteorological surveys should be made when sites are under consideration as development areas for future use by large emitters of gaseous pollutants. It is shown that further general work is required to improve the accuracy of meteorological parameters in equations for plume rise, plume disperson, and time-average ground-level concentrations. An account is given of the wind-tunnel studies made of three coastal sites for new power stations, each site having high ground inland. In addition, reports are given of studies with smoke rockets. A method of avoiding visible dust emission by the installation and effective maintenance of electrostatic precipitators is considered, although the problem of persistent plumes of sulphur trioxide has yet to be solved. The use of dry cooling towers to eliminate the unpleasant clouds of steamy vapour emitted by large stations is discussed, and why this would make it possible to site stations on high ground, thereby ensuring an improved dispersion of the flue gases. A comparison indicates that the reduction of the sulphur content of fuel oils at refineries, with the production of readily stored and saleable sulphur, is to be preferred to those processes which remove sulphur oxides from flue gases.

INTRODUCTION

Choice of sites

POWER STATIONS of 2000-MW capacity are now becoming commonplace. Twelve fossil-fuel-fired stations of this capacity are expected to be in service by 1980, including Drax, 4000 MW, coal.

The main pollutant to which objection is taken is sulphur dioxide (SO_2). An appreciable reduction of the sulphur content of coal is impracticable; and a similar reduction in oil is still too expensive for large power stations. The removal of SO_2 from the flue gases involves adding to a power station a bulky chemical process plant, which, naturally, is resisted by the boiler management. The present method for the disposal of gaseous pollutants is dispersion from tall chimneys, preferably from a single multi-flue structure serving all the boilers.

The rate of emission of SO_2 for each 1 per cent in oil fuel is about 100 t/day per 1000-MW output. In addition, there are emitted about 30 t/day of oxides of nitrogen (expressed as NO), irrespective of the sulphur content of the fuel; these slowly oxidize to nitric acid. Other gaseous pollutants are hydrochloric acid and traces of partially burnt and often unsaturated hydrocarbons. The residual fine dust emitted includes traces of most of the elements in the periodic table, some of which are toxic in high concentrations. This is a vast amount of air contaminant to discharge from a point source, however high the chim-

The MS. of this paper was received at the Institution on 23rd November 1970 and accepted for publication on 17th December 1970.
* *Consulting Chemical Engineer and Fuel Technologist, Woodley, 57 Woodley Lane, Romsey, Hants. SO5 8JR.*

ney. Sulphur dioxide is regarded as the marker for the total pollution from a power station. The capacity of wind currents to dilute the pollutant on all occasions to an innocuous concentration by the time it has reached the ground by eddy diffusion is limited. Conditions on the ground would be worse should the gases be discharged from the boilers through a number of chimneys, even though they were all as tall as a single chimney of a height chosen by the means to be described.

The over-concentration of large sources of air pollution in a few European areas where there is cheap electricity was a matter of concern at a conference in 1964 (1)†. The difficulty has been expressed quantitatively in a recent paper to the U.S. National Air Pollution Control Association. It was shown that the total number of 2000-MW stations that could be permitted in an area of 12×10^4 km^2 in central U.S.A. would be only seven. (The assumptions were: an SO_2 mass emission from other sources equal to that from the power stations, a two-day period of air stagnation, a mean mixing depth of 500 m, an SO_2 half-life of 48 h, and a 24 h air-quality standard of 286 µg/m^3 (s.t.p.) at 0·1 p.p.m. by volume.)

Therefore surveys in depth should now be made by the meteorological services of all industrial countries on the air-pollution dispersion characteristics of those areas where large emitters of potentially harmful gases are likely to be built. To date, nothing has been done, except in respect of nuclear power. The U.S. Atomic Energy Commission has produced (2) 'conservative guide-lines' that may be used

† *References are given in Appendix 61.1.*

in the absence of meteorological experimental data when application is made for a permit to build. This means that explicit data describing the site meteorology are collected during the construction phase. This is too late; history does not record the number of sites abandoned when construction has been well advanced!

Notation

C_3 Ground-level concentration (g.l.c.) of pollutant at the place of interest for a sampling time of 3 min, $\mu g/m^3$

g Gravitational acceleration, 9·81 m/s^2.

H Effective height of plume, m $(h_c + \Delta h - h_e)$.

h_c Height of chimney above base level, m.

h_e Fraction of height of place of interest above chimney base, m.

Q Rate of emission of gaseous pollutant, g/s.

Q_H Heat content of chimney effluent above ambient air, MW.

S $[= (g/T)\, d\theta/dz]$, s^{-2}.

T Absolute temperature of ambient air; taken as 283 K.

u Wind velocity at plume height, m/s.

x Distance along wind from base of chimney, m.

α Parameter in plume rise, equation (61.1), m.

Δh Plume rise, m.

$d\theta/dz$ Mean potential temperature gradient in the atmosphere from the top of the chimney to the top of the plume, degC/m.

σ_y Standard deviation in the cross-wind direction of the plume-concentration distribution, m.

σ_z Standard deviation in the vertical of the plume-concentration distribution, m.

THE PUBLIC ATTITUDE

Whenever proposals are made for an enlargement of the electricity supply service to a nation, the very bulk and height of the buildings for the latest generation of power stations of over 1000-MW output, coupled with many tall cooling towers and one or two chimneys about 200 m high, are sufficient to cause apprehension by the general public. The grounds for objection always include air pollution by sulphur oxides, by clouds of steamy vapour from cooling towers (inland stations), from traffic during construction, and thereafter from traffic when solid fuel is used. In addition, noise has now become recognized as a form of pollution. Loss of visual amenity, particularly in open country districts or by the sea, takes precedence with some objectors. (These remarks apply equally, of course, to proposals for oil refineries, chemical plants, and works for the extraction of metals from ores.) Yet all the objectors want the cheapest possible electricity and an absence of power cuts during inclement weather; therefore large power-production units are essential to obtain the economies of scale. Because boiler houses may be 70 m high they cannot be put underground; and if they are put into a fold between low hills, the chimneys would have to be correspondingly higher.

Visibility of chimney plumes

Public opinion is now demanding invisible, or nearly invisible, chimney plumes. This is particularly so in the U.K. in and around those towns where smokeless fuel has to be used under the 'smoke control' regulations to conform with the Clean Air Act 1956. A smoking chimney is now immediately noticed, whereas a few years ago it would have been observed without comment.

There should be a complete absence of smoke from large modern coal- or oil-fired boilers. The efficiency of dust collection now required and being achieved in the latest U.K. power stations is 99·3 per cent, corresponding with an emission of 0·1 g/m^3, s.t.p. This low concentration is barely visible against most skies. There remain, however, many plumes which are visible over several kilometres travel in certain atmospheric conditions; for instance, when plume widening by atmospheric turbulence is small. These are called persistent plumes. Design information for their prevention when the cause is dust has been given by Jarman (3).

The presence of sulphur trioxide (SO_3) is a cause of persistent plumes from oil-fired boilers. Its formation within the boiler furnaces and passes is inhibited by operating with less than 1 per cent of free oxygen in the combustion chamber. Nevertheless, the problem has remained at one U.K. station with a 130-m chimney where coal-fired boilers were converted to oil firing. The excess SO_3 is probably formed at 500°C in dead spaces installed for the settling of grit. This particular visibility problem has been eased by the injection of fine dolomite powder at the base of the chimney (3). A fully satisfactory explanation of the formation of SO_3 plumes is, however, not yet available. The effluents from many large oil-refinery chimneys are invisible, although the furnaces are not operated at such low excess air.

CHOICE OF CHIMNEY HEIGHT

There are still a few unknowns in the calculation of the ground-level concentration (g.l.c.) of pollutants from a high chimney. The flue-gas plume rises so high that the highest g.l.c. from a 2000-MW station normally occurs at a distance of around 10 km. At this distance the ground around may be higher than that of plant base level, and even higher than the chimney. Moreover, plants near towns must now have chimneys designed to take into account the existence of, or the possibility of, the construction of tall buildings of 100-m height (more in the U.S.A.). For these reasons it is most essential to conduct wind-tunnel studies with a landscape model to determine the minimum required height of a particular chimney. This height should be compared with experience gained elsewhere. Whenever possible, smoke trails should then be laid by a helicopter or a slow-flying aeroplane at heights around the estimated plume height when the wind is blowing at various speeds over those parts of the landscape indicated as critical by the wind-tunnel tests.

As a final check, calculation should be made of the g.l.c.,

Source: Central Electricity Generating Board

Fig. 61.1. Multi-flue 200-m-high chimney for Eggborough 2000-MW power station

using the latest established methods, and for various weather conditions. The highest g.l.c. so obtained must be below any time-average limits set by government or local authorities.

WIND-TUNNEL STUDIES

The author has been associated with the studies of plume travel from three proposed coastal power stations with hilly landscapes inland. Two of these were in wind-tunnels in the National Physical Laboratory (N.P.L.) and the third was at Bristol University.

New Plymouth, New Zealand

The proposal was for a 600-MW coal-fired power station. The proposed site is on the east coast of North Island, facing north with the town to the west. The ground rises gradually in 17 km to 1200 m, and then steeply over the next 8 km to the summit of Mount Egmont, which is 2500 m. There is an arboretum of special rhododendrons at a distance of 16 km, height 490 m. The inhabited area close to the station comprises the town at sea-level, and

outlying groups of villas at 60–90 m. Immediately adjacent to the proposed chimney is a steep, conical hill, Paritutu, 153 m high. There would be a single five-flue chimney serving the five boilers. Balloon and smoke-rocket tests (4) by the New Zealand Meteorological Service on site showed that the disturbance of air flow above Paritutu extended to a height of only 30 m. Hence, most of the wind-tunnel tests were made with a chimney 183 m high.

The diameter of the top of the windshield was 18 m, and the flues projected above it by 6 m (by one-third of its diameter) in order to reduce downwash. This is an arrangement found in wind-tunnel tests by the British Central Electricity Generating Board (C.E.G.B.) to be effective. Fig. 61.1 is a photograph of the arrangement at Eggborough.

Calculations using the Lucas–Pasquill equations given later showed that the wind speed giving the highest g.l.c. was 10 m/s, a speed also found by measurement at High Marnham, having two chimneys of 137 m, 1000-MW capacity (5). As there are frequent winds of greater speed at the site, and as it was important to avoid impingement of the plume on the vegetation and footpaths on Paritutu,

Source: New Zealand Electricity Dept

The left side of the tunnel is open for observation. The model is of a proposed coastal power station at New Plymouth, New Zealand. The height of the conical hill near the chimney is 150 m. Consulting engineers: Preece Cardew & Rider, Brighton, England. Crown Copyright reserved.

Fig. 61.2. Landscape model (1/1760) on turntable in open sided wind-tunnel, National Physical Laboratory, Teddington

tests were conducted at speeds of 10 and 15 m/s. It was determined that downwash occurred, and so there was a possibility of plume impingement, when the effluent velocity from the flues was less than 23 m/s. It was therefore decided to design for 26 m/s at full load. A higher velocity would have been inadvisable because there would be a positive pressure at the chimney base should the velocity be increased to 29 m/s.

Fig. 61.2 shows the model mounted on a 2·4-m diameter turntable in the N.P.L. tunnel. The effective length of the tunnel is 3·65 m. The scale of the model is 1/1760. Each band on the marker posts represents 33 m.

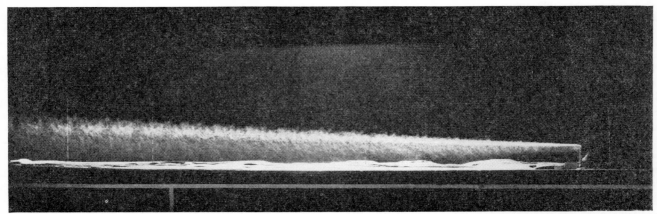

Source: New Zealand Electricity Dept

Chimney height 183 m, diameter 15·3 m, efflux velocity 23 m/s, wind speed 15 m/s. Downwash is occurring because the efflux velocity is too low for such a wide chimney.

Fig. 61.3. View of model in Fig. 61.2 showing plume flow integrated over 2 s

Source: South of Scotland Electricity Board

Fig. 61.4. View across the Firth of Clyde showing Inverkip power station (2640 MW) as it will appear against the background of hills

To simulate the plume rise from the heat emission of 75 MW in the flue gases the smoke in the model was carried in helium. The air speed and the effluent velocity were adjusted to give the same ratio of momentum per second per unit area as would apply on the full scale. Turbulence in the empty tunnel is 1 per cent but is increased appreciably at low level to about the figure of 3 per cent found over the open sea. Roughness over the full-scale landscape derived from houses and trees was simulated with corrugated cardboard.

Fig. 61.3 shows a photograph of the plume passing over the model: wind speed was 15 m/s; effluent, 23 m/s. Some downwash can be seen. With an effluent speed of 15 m/s the downwash extended half-way down the side of the chimney. The photograph was taken with stroboscopic illumination, giving 20 flashes during 2 s.

A few smoke trails laid by the New Zealand Air Force did not indicate any effects not seen in the wind tunnel. The tests were insufficient in number to be unquestionably confirmatory. Nevertheless, it was a useful precedent for the defence forces of a nation to assist in combating air pollution.

It was concluded from the tests, and by calculations of the highest g.l.c. for the worst weather conditions, that a chimney of 183 m height would be sufficient. To provide a factor of safety for Paritutu, the Chief Chemical Inspector of the New Zealand Health Department has stipulated a chimney of 213 m.

There remains some doubt concerning the effect of frequent cold-air streams (katabatic winds) blowing down during the night from the heights of Mount Egmont towards New Plymouth. Thirty-five per cent of these winds are between 150 and 300 m deep. If the upper wind is northerly, the plume will be carried inland above the katabatic wind and a part might be entrained on the mountain slope and be carried back to the coast at ground-level. This point could not be studied in a wind-tunnel. In the absence of studies of air flow on site with suitable tracers (**6**), differing opinions were obtained from meteoro-logists as to the possible magnitude of the effect. The subject is mentioned here because there are likely to be many large power stations where it will be essential to take into account the effect of distant mountains.

Wanganui, New Zealand

This site was considered as an alternative to New Plymouth. In the wind-tunnel tests the plume from a 175-m chimney dipped as it passed over a river estuary bounded by hills of 30–70 m. The dip did not occur when the model chimney was increased to 185 m. The fact that a similar dip of the air currents was observed on site with balloons gave confidence in the wind-tunnel observations.

Inverkip, Scotland

This will be a 1980-MW, oil-fired station, with an ultimate capacity of 2640 MW, situated on the east shore of the Firth of Clyde. The site is backed by hills rising to 300 m at 4 km, and then to 425 m at 10 km distance. To the north-east there is a village at 7 km distance and 150 m height, situated by a steep valley. The village lies at the distance of the calculated highest g.l.c. Across the Firth (to the west) there are hills of 400 m at a distance of 6 km, which could affect the wind flow over the station. Fig. 61.4 shows the four-flue chimney of the chosen height of 240 m superimposed on a photograph of the site.

A model of the site to the scale of 1/1000 was studied at Bristol University in a closed wind-tunnel 10 m long, 5·5 m wide, and 2·3 m high; a helium plume was also used. Samples of the tunnel air are withdrawn continuously from points on the model to a recording mass spectro-meter and analysed for helium, thus giving records of g.l.c. averaged over a few minutes. Fig. 61.5 shows the smoke plume from a 240-m chimney with an emission velocity of 15 m/s into a wind of 10 m/s. A coloured ciné-film record of the smoke flow was taken to permit repeated study of the flow over the landscape. An uplift of the plume by the 150-m ridge upwind of the village was clearly seen.

Source: South of Scotland Electricity Board

Chimney height 240 m, efflux velocity 15 m/s, wind 10 m/s. At top right there is a village at 7 km distance and 150 m above sea-level. Hills to right (west) rise to 175 m and to 440 m to the south-west.

Fig. 61.5. Wind-tunnel model (1/1000) at Bristol University of Inverkip power station

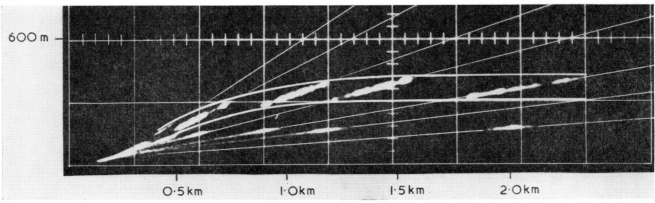

Source: Central Electricity Research Laboratories, Leatherhead

The continuous white lines enclose the outline of the B plume. For additional photophraphs, see reference (7).

Fig. 61.6. Lidar scan of the two plumes from Tilbury power station (360 MW)

In making calculations of the g.l.c. there remained the meteorological problem of high inversions. These are to be expected when warm winds from the Azores are cooled at low level by the Atlantic Ocean, leaving warmer air aloft—an inversion. No measurements were available of the height, depth, temperature gradients, and duration of these inversions. Lidar can be used to detect inversions (7). (Lidar means 'Light detection and ranging' using radar techniques with a pulsed laser beam.)

Site tests with rockets

Wind-tunnel tests are expensive and may take over a year to arrange. Kelly (4) describes the successful use of inexpensive 'smoke-puff' rockets to assist in the choice of a height for a chimney in Ireland situated beside a 700-m-wide river running past hills which rise to 85 m. The rockets were fired at an angle to give the estimated height of plume. The tests showed the position and intensity of the downward currents of air caused by the hills, and they indicated the minimum height of chimney required to prevent impingement of flue gases on the hills.

Better information could have been obtained, but at much greater expense, with smoke trails laid by slow-flying aircraft during those weather conditions which would give the highest g.l.c. The method is not regarded as practicable for the U.K. and Western Europe because of the unpredictability of the weather conditions.

PLUME RISE

Dispersion depends, in part, on the height to which the plume of hot flue gas rises by virtue of its buoyancy. We now have two reasonably accurate but empirical formulae for estimating the plume rise from a chimney higher than 100 m serving a station emitting up to 100 MW of heat. The first is that derived by the C.E.G.B. research laboratory using pulsed laser beams to track the path of the plume. Fig. 61.6 shows a lidar trace of one of the two plumes from Tilbury power station on the Thames estuary (7). The plume is seen to continue to rise for more than 2 km.

The formula proposed by Lucas (8) for plume rise averaged over 1 h for a heat emission of 4–70 MW to a neutral atmosphere from chimneys of height 60–130 m is:

$$\Delta h = \alpha Q_H^{1/4} u^{-1} \quad . \quad . \quad . \quad (61.1)$$

where

$$\alpha = 475 + \tfrac{2}{3}(h_c - 100) \quad . \quad . \quad (61.2)$$

Lucas believes that α may increase substantially with further increase in chimney height, e.g. the factor 2/3 may become greater than 2.

Equations (61.1) and (61.2) are used in A.S.M.E. specifications (9) APS 1 and APS 2.

Alternatives to equations (61.1) and (61.2) have since been published by Briggs (10) after a survey of all the data available, including plume rise measured in the U.S.A. by SO_2 probes suspended from helicopters, and by infrared photography. For near neutral and unstable conditions (categories C/D and D defined in Table 61.1), and for heat emissions of 5–90 MW, the Briggs formula simplifies to:

$$\Delta h = 3 \cdot 3 Q_H^{1/3} u^{-1} (x_f)^{2/3} \quad . \quad . \quad (61.3)$$

where x_f is the distance downwind at which the plume rise begins to level out; this is approximately the distance of the point of the highest g.l.c. For large heat emissions Briggs suggests $x_f = 10 h_c$, although Fig. 61.6 shows that plumes may continue to rise for double this distance. Further investigation is required to define the value of x_f to be used for smaller and larger heat emissions.

For stable conditions (categories D/E–F) at the heights covered by the plume travel, the average final height is given by Briggs as:

$$\Delta h = 5 \cdot 9 Q_H^{1/3} (uS)^{-1/3} \quad . \quad . \quad (61.4)$$

where $d\theta$ is in degC and dz in metres.

The value suggested for $d\theta/dz$ for stable conditions is 9 degC/km, but it is better to obtain meteorological advice for each site. The number of measurements used for the derivation of equation (61.4) is limited, and further work is required.

Table 61.1. Key to Pasquill stability categories (14) (15)

Surface wind speed (at 10 m), m/s	Strong	Insolation		Night-time conditions (including 1 h before sunset and after dawn)	
		Moderate	Slight	Thin overcast or 4/8 low cloud	3/8 cloud
2	A	A–B	B	—	
2–3	A–B	B	C	E	F
3–5	B	B–C	C	D	E
5–6	C	C–D	D	D	D
>6	C	D	D	D	D

A = extremely unstable conditions; B = moderately unstable conditions; C = slightly unstable conditions; D = neutral conditions and for all overcast conditions during day or night, irrespective of wind speed and for any sky condition during the hour preceding or following night; E = slightly stable conditions; F = moderately stable conditions.
  For A–B take averages of sigmas for A and B, etc.
  Strong incoming solar radiation corresponds to a sun altitude greater than 60° with clear skies; slight insolation to a sun altitude from 15 to 35° with clear skies. Incoming radiation that would be strong with clear skies can be expected to be reduced to moderate with 5/8 to 7/8 cloud cover of middle height and to slight with broken clouds.
  Frequencies of categories in Britain can be obtained from the Meteorological Office.

Preliminary measurements around the larger (150 MWh) T.V.A. plants indicate that the frequency and magnitude of the g.l.c. are higher than with the simpler case for neutral conditions (**11**). In other words, the presence of frequent elevated inversions of temperature gradient in the upper atmosphere may become a controlling factor in the estimation of the g.l.c. More intensive studies of the subject in Europe are indicated, and in the industrial coastal districts of South Africa and Australasia also.

Multiple separate chimneys

A survey by Briggs showed little or no enhancement in neutral conditions of plume rise when an adjacent chimney of about the same heat emission was commissioned.

PLUME FLOW OVER HILLY LANDSCAPE

Some deduction from the plume height, $H = h_c + \Delta h$, is necessary in calculations of g.l.c. when the landscape around a station is higher than the chimney base (see above,

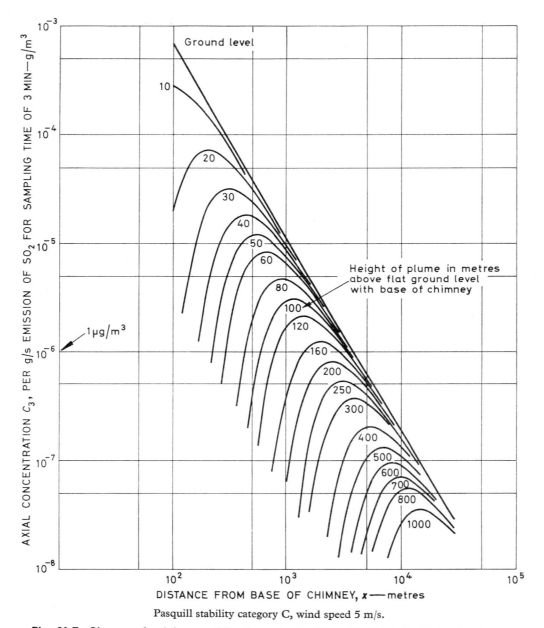

Pasquill stability category C, wind speed 5 m/s.

Fig. 61.7. Change of axial ground-level concentration with distance from source for various plume heights (m)

'Wind-tunnel studies'). This is a subject on which no advice is given in summary works such as references (6) and (9)–(13).

Frankenberg (12) reported on observations in a wind-tunnel showing a 'ski-jump' effect of 15–45 m over a steeply rising 100-m-high ridge downwind of Clifty Creek when the chimney was at a sufficient height to ensure that the entrainment of the plume did not occur in the tur-bulent boundary layer of about 50 m thickness above the ridge. Hamilton (7) reported that lidar observations of the Northfleet plume indicated a fall of 30 m in plume height at high wind speeds, which were caused by wind blowing off a cliff 30 m high, 2000 m long across wind, and 500 m upwind of the chimney.

In the absence of wind-tunnel tests the author suggests for initial calculations that half the height of a ridge longer than 1000 m be subtracted from the plume height when the angle of incidence is 45–90°, and that the full height be subtracted when the incidence is less than 45°. The case of an isolated hill or tall building is quite different because the wind flow merely divides round the obstacle. The concentration to be calculated is that within the plume at the height and distance concerned (14).

When there is a ridge at a distance of less than 2000 m upwind of the station, allowance for the height of the ridge above the chimney base must be subtracted from the plume height. This correction is frequently overlooked.

The above suggestions can apply only to hills, not mountains.

PLUME DISPERSION

The equations for the dispersion of a plume are now well established. They are a development of the well-known 1947 Sutton equations by Pasquill (15), as rearranged by Gifford (16). A valuable 'workbook' for design engineers which presents these equations with worked examples has been issued by the U.S. Dept. of Health (14). The shape of the curves relating the maximum 3-min g.l.c. from plumes of various heights is shown in Fig. 61.7.

In the main, we are concerned with the circle of the highest g.l.c. around a station. The equation then simpli-fies to:

$$C_3 = (0.117 \times 10^6 Q)/(\sigma_y \sigma_z u) \quad . \quad . \quad (61.5)$$

where C_3 is the maximum 3-min g.l.c. of pollutant in $\mu g/m^3$ at ambient temperature. The values of σ_y and σ_z are given in Figs 61.8 and 61.9 for the six atmospheric stability categories suggested by Pasquill and defined in Table 61.1. The original values of the sigmas given by Pasquill have been confirmed by Kawabata (17) up to distances of 5 km. Beyond this distance the values are tentative, and further data need to be collected by meteorologists for the industrial areas of every continent. In addition, the frequency and duration of the occurrence of the stability categories should be listed.

For the highest g.l.c.

$$\sigma_z = H/\sqrt{2} \quad . \quad . \quad . \quad . \quad (61.6)$$

The distance x of the highest g.l.c. is read from Fig. 61.9, and then the value of σ_y at x is read from Fig. 61.8.

For estimation of the highest g.l.c. when there are inversions which will not be penetrated by the rising plume, reference should be made to the 'workbook' (14). The highest calculated g.l.c. obtained from equation (61.5) is increased up to double its value if the upper edge of the expanding plume is reflected downwards from an elevated inversion, as discussed by Scriven (18).

The penetration of two hot plumes from the 90-m chimneys of Brunswick Wharf power station, London, through fog over London extending to a height of 250 m in 1962 is illustrated in Professor Scorer's book of air-pollution photographs (19).

TIME-AVERAGE CONCENTRATIONS

Equation (61.5) gives the maximum concentration of pol-lutant during any time-sample of 3 min, and during any hour the value of C_3 may vary between zero and the maximum. Since the wind varies in direction and speed, and the variations have a complicated pattern in frequency and amplitude, particularly close to the ground, the g.l.c. varies greatly from moment to moment. Lucas (8) states that the concentration pattern only becomes fairly steady when averaged over an hour. It is, in fact, becoming common practice for pollution specialists to use the 1-h average g.l.c. of SO_2 as the significant parameter in con-sidering the magnitude of air pollution from a source [e.g. by A.S.M.E. (13)]. In the main, health authorities use measured 24-h average concentrations of SO_2, because this is the measurement made throughout the year in many towns, particularly in the U.K. These records measure the 'background' pollution derived from all points of the compass.

Hence, the ratios 1 h/3 min and 24-h/3 min are the most important when considering the significance of the g.l.c. from a new or existing chimney. The time-average g.l.c. is added to the background SO_2 when considering the effects on health of a new station.

A comprehensive review of the literature on the effects of SO_2 on man, animals, vegetation, and materials is given in reference (20). The standards for time-average per-mitted additions to the ambient SO_2 concentration set by local and government authorities vary widely. One of the strictest is that set by the city of Amsterdam for the addition of SO_2 from the chimney of a new oil refinery. This requires not more than 100 $\mu g/m^3$ over 24 h.

It follows from the above discussion that the ratios 1 h/3 min and 24 h/3 min g.l.c. are the most important. These ratios are dependent on variations with time of the ratio σ_z/σ_y and the wind speed. It is implicit that the wind direction remains constant within the usual 30° sector commonly employed in meteorological records. There are insufficient records of the variations of σ_z/σ_y with time for time factors to be derived from local meteorological in-formation, and it is necessary to use available records of 3-min maximum g.l.c. of SO_2 over prolonged surveys. Analysis of the highly complicated records taken during

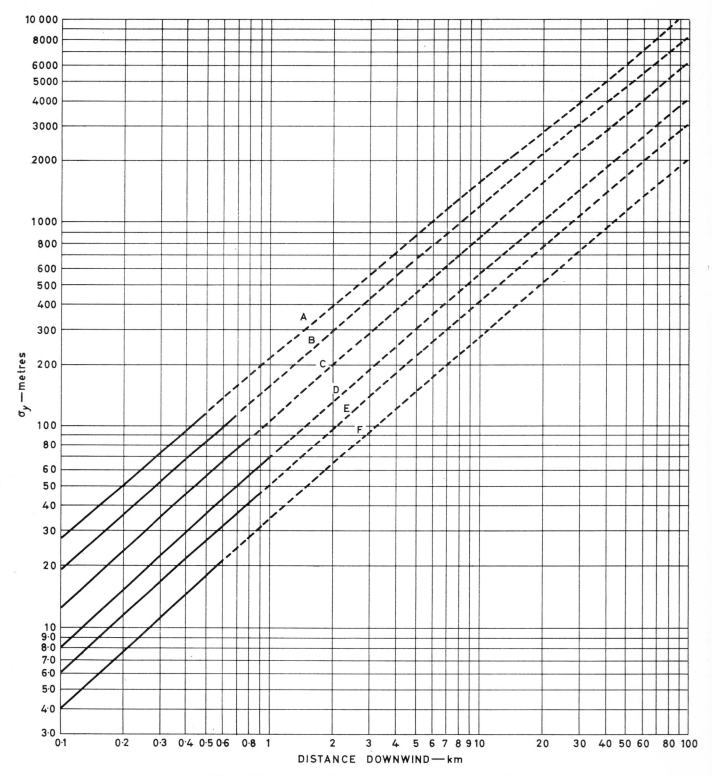

Derived from reference (**15**), this version from reference (**14**).

Fig. 61.8. Horizontal dispersion coefficient σ_y as a function of downwind distance from the source

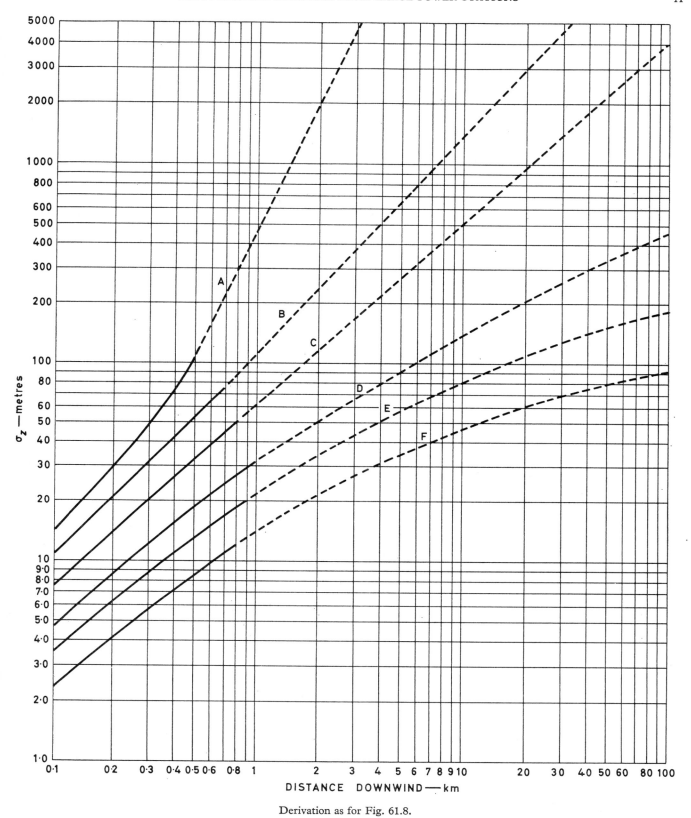

Derivation as for Fig. 61.8.

Fig. 61.9. Vertical dispersion coefficient σ_z as a function of distance from the source

Source: E.E.–A.E.I. Turbine Generators Ltd, and C.E.G.B.

Fig. 61.10. Natural-draught dry-cooling tower, Rugeley

the Tilbury survey (**21**) indicates that the 1 h/3 min ratios can be simplified to the following:

Steady wind conditions	3/4
Average conditions	1/2
Certain unstable conditions	1/4

For the highly unstable conditions, category A, Maynard Smith (**22**) suggests that the ratio may be 15. This would only apply for a 1–2 m/s wind. For most power station cases the highest g.l.c. are found when the stability is C and the wind 8–12 m/s, and the ratio may be 3/8 (average of 1/2 and 1/4).

For the ratio 24 h/3 min the factors 1/10 and 1/12 were found at High Marnham (**5**) and Tilbury (**21**). The conservative ratio 1/10 seems appropriate.

For further discussion of time-averages, see Hino (**23**).

THE PROBLEM OF COOLING TOWERS

The amount of water vapour discharged from wet-type cooling towers serving a 2000-MW station is more than 3000 t/h. There is a group of eight towers at least 100 m high. Effective means of preventing droplet carry-over are now installed.

In light winds and high humidity a plume will persist for several kilometres, having the appearance of stratus cloud. This is normally accompanied by a stable atmosphere and a sky which is overcast. With light winds and an unstable atmosphere a cumulus cloud may form above the towers. Sometimes this occurs without other clouds in the sky, but the clouds would probably have formed later.

Since it is not possible to build all power stations on the sea-coast, and since flue gas dispersed in hilly country would be better if the chimneys were on high ground,

Table 61.2. Dimensions of natural-draught air-cooling tower, C.E.G.B., Rugeley

Items	Details
Tower—	
Height	109 m
Diameter, base . . .	100 m
Diameter, throat . .	62 m
Wall thickness . . .	125 mm
Draught provided . .	0·5 mb
Cooling elements—	
Height	14·6 m
Surface area . . .	35 000 m²
Frontal area . . .	7500 m²
Material . . .	aluminium
Water circulation rate . . .	47 m³/s
Fresh water saved per day . .	7000 m³

there was a challenge to inventive engineers to reduce the requirement of water and the unpleasant emission of vapour-mist by the application of air-cooling technology. Use of air coolers is now standard practice in modern oil refineries, but the temperature requirement is easier.

A full-scale prototype natural-draught parabolic tower air cooler has been in successful operation since 1962, serving a 120-MW base-load generator at the C.E.G.B. power station at Rugeley in central England. The design is based on that proposed by Professor Heller, Budapest, at the World Power Conference, 1956. Fig. 61.10 shows the tower in operation with water-mist plumes rising from conventional water-cooling towers serving other genera-tors. Relevant technical details are given in Table 61.2. The tower is substantially larger than a wet-cooling tower for the same duty.

Initially, there was some crevice and deposit corrosion of the coolers caused by the sea-salt and smoke content of the English air. This has been overcome by coating the vulnerable areas with a 25–30 μm thick film of epoxy resin. The operational design has successfully prevented freezing troubles at −10°C.

Full details of the initial development work, of the design and economics of the system, and of the teething troubles have recently been published by the Institution of Mechanical Engineers (24). With the experience gained, Heller system plants are in operation in the German Federal Republic (150 MW), Hungary (200 MW), and the Republic of South Africa (200 MW). A plant is being built in Russia to serve 3 × 220 MW generators.

DISCUSSION

The information summarized in this paper has been used by the author for the preparation of evidence given at public inquiries in Britain on the effects of pollutants from the chimneys of proposed large projects. These practical exercises have shown up difficulties in the interpretation and application of equations for the dispersion of gases from tall chimneys. Much more additional meteorological data are required. It is highly desirable that this informa-tion should be acquired during the period when a search is

being made for sites for any process plants that will emit massive quantities of gaseous pollutants. It must be admitted that a meteorological survey of the pollution dispersion characteristics will take more than two years to arrange and complete in all essentials.

Improvements are required in the techniques of wind-tunnel examination of the flow of gases over hilly land-scapes.

It is hoped that this paper will assist project engineers in an appreciation of their responsibilities in respect of air pollution and in discussions with meteorologists. It may be found that solutions to some of the problems mentioned have already been obtained by experience, possibly in other countries.

Problems of dust emission have not been discussed, since dust emissions can now be reduced to innocuous quantities by careful attention to the design and continuous maintenance of dust arrestors—but the cost is high (25).

A review of the costs of removing sulphur dioxide from power-station flue gases has been given recently by Billinge (26), and of the cost of reducing the sulphur con-tent of oil by Fuller (27). In the author's opinion, reduction of the sulpher content of residual oils during the process of refining will eventually be the economically acceptable solution of the problem of sulphur dioxide emissions from 2000-MW stations which are sufficiently close together for their effluents to affect the environment. Much of the publicity on the adverse effects of sulphur dioxide from such stations is ill-informed (28).

ACKNOWLEDGEMENTS

The author is indebted to many of the authors named in the list of references for fruitful discussions on the problems of air pollution during the past 35 years, and to those who have provided the illustrations for this paper. The treatment of dispersion is based on that in the revised edition of *Gas purification processes* (Butterworths), in the press.

APPENDIX 61.1
REFERENCES

(1) Report on European conference in 1964 on air pollution, Strasbourg', *Int. J. Air Wat. Pollut.* 1965 **9**, 11.

(2) NIEMEYER, L. E., McCORMICK, R. A. and LUDWIG, J. H. 'Environmental aspects of power plants', *Symp. Int. Atomic Energy Agency*, New York, 1970.

(3) JARMAN, R. T. and DE TURVILLE, C. M. 'Visibility and length of chimney plumes', *Atmosph. Envir.* 1969 **3**, 257. Report of discussion of this and other papers, *Atmosph. Envir.* 1970 **4**.

(4) KELLY, A. G. 'Smoke rockets to establish chimney heights', *Engr Bldr* (House Review), 1966 (April), 102.

(5) MARTIN, A. and BARBER, F. R. 'Investigations of sulphur dioxide pollution around a modern power station', *J. Inst. Fuel* 1966 **39**, 294; *Atmosph. Envir.* 1967 **1**, 655.

(6) SLADE, D. H. 'Meteorology and atomic energy', TID-24190, 1968 (U.S. Clearing House for Technical Information, Springfield, Va. 22151, U.S.A.).

(7) HAMILTON, P. M. 'The application of a pulsed-light range-finder (Lidar) to the study of chimney plumes', *Phil. Trans. R. Soc.*, Series A, 1969 **265**, 139.

(8) LUCAS, D. H. 'Symposium on plume rise and dispersion', *Atmosph. Envir.* 1967 **1**, 351.

(9) *Recommended guide for the control of emission of oxides of sulphur* 1970 (Am. Soc. Mech. Engrs, New York, N.Y., U.S.A.).

(10) BRIGGS, G. A. 'Optimum formulae for buoyant plume rise', *Phil. Trans. R. Soc., Series A*, 1969 **265**, 197; 'Plume rise', TID-25076, 1969 (U.S. Clearing House for Technical Information, Springfield, Va. 22151, U.S.A.).

(11) CARPENTER, S. A. *et al.* 'Plume dispersion models, T.V.A. plants', *Annual Meeting Air Pollution Control Assoc.* 1970.

(12) SPORN, P. and FRANKENBERG, T. T. 'Pioneering experience with high stacks', *Proc. Int. Clean Air Congress* 1966 (Nat. Soc. Clean Air, London).

(13) *Recommended guide for the prediction of dispersion of airborne effluents* 1968 (Am. Soc. Mech. Engrs, New York, N.Y., U.S.A.).

(14) TURNER, D. B. 'Workbook of atmospheric dispersion estimates', U.S. Public Health Service Publication No. 999-AP-26, 1969 (Cincinnati, Ohio, 45227, U.S.A.).

(15) PASQUILL, F. 'The estimation of the dispersion of windborne material', *Met. Mag.* 1961 **90** (No. 1063), 33; *Atmospheric diffusion* 1962 (van Nostrand, London).

(16) GIFFORD, F. A. 'Use of routine meteorological observations for estimating atmospheric dispersion', *Nucl. Saf.* 1961 **2** (No. 4), 47.

(17) KAWABATA, Y. 'Observations of atmospheric dispersion', *Geophys. Mag. Tokyo* 1960 **29**, 571.

(18) SCRIVEN, R. A. 'Properties of the maximum ground-level concentration from an elevated source', *Atmosph. Envir.* 1967 **1**, 411; *Phil. Trans. R. Soc., Series A*, 1969 **265**, 209.

(19) SCORER, R. S. *Air pollution* 1968 (Pergamon Press, Oxford).

(20) 'Air quality criteria for sulphur oxides', U.S. National Air Pollution Control Administration Publication No. A.P.-50, 1969 (Washington, D.C., U.S.A.).

(21) LUCAS, D. H. Private communication, 1970.

(22) SMITH, MAYNARD E. 'Discussion on proposed Canadian standard', *Atmosph. Envir.* 1969 **3**, 487.

(23) HINO, M. 'Maximum ground-level concentration and sampling time', *Atmosph. Envir.* 1968 **2**, 149.

(24) CHRISTOPHER, P. J. and FORSTER, V. T. 'Rugeley dry-cooling tower system', *Proc. Instn mech. Engrs* 1969–70 **184** (Pt 1, No. 11), 197.

(25) LOWE, H. J. 'Recent advances in electrostatic precipitators', *Phil. Trans. R. Soc., Series A*, 1969 **265**, 301; and Chap. 13D, *Gas purification processes* (ed. G. Nonhebel), 2nd edn, in prepn, 1971 (Butterworths, London).

(26) BILLINGE *et al.* 'Reduction of emission of sulphur dioxide', *Phil. Trans. R. Soc., Series A* 1969 **257**, 309.

(27) FULLER, H. I. 'Sulphur oxides and oil fuels', *Proc. Clean Air Conf.* 1969 (Nat. Soc. Clean Air, London).

(28) *One hundred and sixth Annual Report on Alkali, etc., Work* 1969 (H.M.S.O., London).

C64/71 THE CONTRIBUTION OF SERVICE EXPERIENCE TO THE DEVELOPMENT OF MODERN LARGE STEAM TURBINES

N. C. PARSONS*

The development of modern, large steam turbines is backed not only by detailed analytical design and laboratory research and development but also by intelligent interpretation of service experience and measurements taken on operating machines. The paper discusses the latter aspect and describes a number of relevant examples.

INTRODUCTION

FROM THE INCEPTION of the steam turbine, development of improved and larger machines has included a thorough analysis of experience accumulated on machines in service. The quest for higher efficiency, greater loading of materials with a view to economy of the product, and deeper understanding of design fundamentals has led to growing investment in development and research carried out in laboratories, using test rigs such as model turbines, full-scale blade-vibration facilities, large bearing rigs, erosion testers, and so on. The results obtained from such work have been invaluable in developing efficient and reliable designs for the ever-larger turbines that are a feature of the rapid growth of the electricity-supply industry.

Furthermore, during the past decade there has been a marked increase in the use of analytical methods for design, as a result of the combination of a rapid advance in technology and the availability of more elaborate and faster computers, which have made practicable complex calculations that were formerly impossible to carry out by hand. The existence of computers has also encouraged the development of more advanced methods of analysis, so the extent to which analytical design is employed is still an accelerating trend.

In spite of this advance there still remain areas of design where the technology is not sufficiently well understood and the analysis is not sufficiently developed to reveal potential difficulty or trouble. While in some areas it remains possible to obtain the essential data from well-conceived research in the laboratory, or by using appropriate test-rigs, there are a number of instances where it is not possible to examine the behaviour or performance of components without building full-scale models to represent completely the conditions encountered in service operation. As this latter course would be prohibitively expensive, the engineering of such features has to be carried out as completely as possible using all relevant experimental and analytical information. It is also essential to feed in experience gained on machines in the field, obtained either from measurements taken on actual machines or thorough analysis of operational difficulties encountered.

This paper discusses the contribution of such field experience, taking as illustrations a number of examples drawn from large steam turbines manufactured by the author's company.

Classification of information

Any classification of information obtained from service experience tends to an extent to be artificial, since a very considerable degree of overlap is bound to exist. For the purpose of this paper, however, the examples chosen could be allocated broadly to the following categories:

(1) Features where the actual operating stress level is largely incalculable, and consequently it is necessary to adopt a design approach that contains a degree of empiricism.

(2) Material problems or deficiencies only revealed during the course of service operation.

(3) Behaviour associated with thermal stressing as a result of transient temperature variations.

(4) Variations in degree of fixity occurring under service conditions.

(5) Investigations or measurements carried out on machines to study phenomena which can only be accurately represented in full-scale operation up to full-power output.

The MS. of this paper was received at the Institution on 1st October 1970 and accepted for publication on 12th November 1970. 22
* Director of Engineering, Turbine and Generator Division, C. A. Parsons & Co. Ltd, Heaton Works, Shields Road, Newcastle upon Tyne, NE6 2YL.

HIGH-PRESSURE CONTROL VALVES
Strength of valve stem

This example refers primarily to item (1) in the above classification, but contains a number of facets which reflect on other of the categorized items.

It is general practice for high-pressure (h.p.) steam-control valves to be of the single-seated, high-velocity type, comprising a suitably profiled head on the valve in conjunction with a valve seat followed by a diffusing section. For many years the arrangement used by the author's company was broadly as shown in Fig. 64.1. Such a valve possesses good control characteristics with a high degree of stability. It has long been recognized that, in addition to the steady forces tending to open or close the valve under any operating condition, there will be vibratory forces due to the non-steady nature of the steam flow. These latter forces are quite incalculable, and attempts to measure them on model rigs in the laboratory have proved of limited use, since it is impractical to reproduce the large mass flow and high Reynolds numbers that exist in service. In fact, up to 1962 the author's company had enjoyed complete immunity from troubles arising from steam-valve buffeting, and this seemed to indicate that the design was not near any condition of danger.

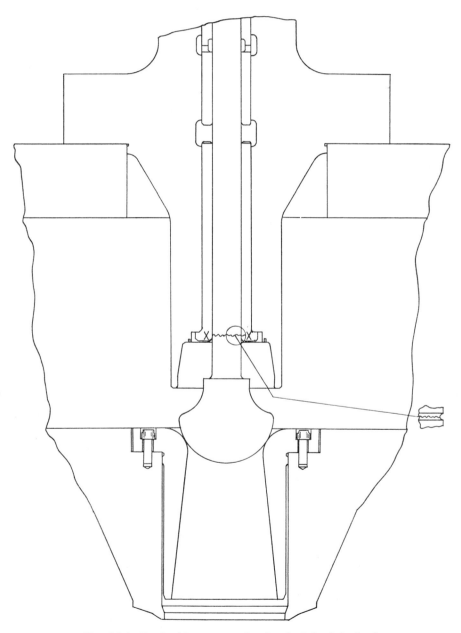

Fig. 64.1. Typical h.p. control valve (original design)

In 1962 a 500-MW machine was commissioned with stop-valve steam conditions of 2400 lb/in² (gauge), 1050°F. This represented a considerable advance in stop-valve mass flow associated with a high inlet temperature. In accordance with practice current at the time, the valve stems had circumferential grooves along their length of engagement in the glands. The advantages claimed for the grooves are: that they provide places where small pieces of scale, etc., can rest harmlessly without scoring the bush; and that they act as pressure-equalizing channels, and so minimize the tendency for the valve stem to bear hard against one side of the bush due to the action of steam pressure. In view of the satisfactory performance on such valves in the past, the action of the grooves as stress raisers was not considered to be significant at the calculated levels of stress.

The situation at the time was further complicated by recent research work (I)* having indicated that nitrided

*References are given in Appendix 64.1.

steel previously used for h.p. valves had a tendency to grow, owing to surface oxidation at temperatures above 1000°F. Since such growth could lead to seizing of the valve spindle in the guide bushes, it became general practice throughout the industry to manufacture valve stems from materials having greater oxidation resistance. In the case being described, it was decided to deposit by welding a thin surface layer of stellite. In the course of manufacture, however, some of the valves were found to be thermally unstable and had to be replaced by valves made in 12 per cent chromium alloy steel without stellite facing.

After the machine had been in service for about 200 h, under somewhat adverse operating conditions so far as the valves were concerned, two of the valve heads—one associated with a stellited stem and the other with an unstellited stem—broke off at the pressure-equalizing groove nearest the head. The fractures were of a fatigue nature and had the appearance of having been caused by

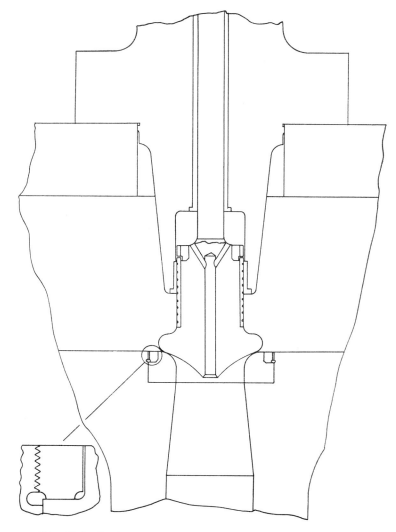

Fig. 64.2. Typical h.p. control valve for large mass flow

alternating bending stresses; almost certainly the result of steam-buffeting forces on the valve head. The obvious solution was to thicken the valve stem, remove the pressure-equalizing grooves, and to use for all the valves a material which possessed adequate resistance to oxidation without requiring stellite surfacing. It is encouraging to note that these modifications have proved entirely successful in that no valve troubles have been encountered since on the particular machine.

Immediately this failure had taken place it was decided to apply the same modifications to the h.p. governor valves for another machine of very similar rating. In the event, however, service requirements made it impractical to carry out the full modification. Instead, only plain, ungrooved valve stems of the original diameter were fitted. As fatigue tests on half-size models of the grooved and ungrooved valve stems had confirmed that the stress-concentration factor of the groove was approximately 3, and therefore in close agreement with published data (2), it was expected that the arrangement would be adequately strong in fatigue. However, after some 10 000 h of service operation one valve head broke off. The steam chest was modified to incorporate the thicker valve stem, and subsequent operation has been entirely satisfactory.

As a result of these two examples, and taking into account the considerable experience accumulated on earlier machines, it has been possible to evolve a satisfactory empirical approach to the design of valves of this type.

Some additional features have also been added which laboratory testing has shown to be beneficial; but owing to the difficulties mentioned above, it is difficult to quantify the effects.

As ratings of machines increase, it becomes impractical to continue to use this design of valve because the diameter and lift increase to such an extent that the bending moment in the valve stem, as a result of the non-steady forces, demands an excessively large diameter of stem. For larger valves, therefore, a guided head design has been adopted, as shown in Fig. 64.2. With this arrangement, the valve stem is protected from excessive bending stress by the large-diameter guiding bush, which can be positioned close to the valve head.

While on the subject of valve-stem problems, it is interesting to record that valves with stems made from nitrided steel were fitted to a machine of smaller rating having a stop-valve temperature of 1050°F. These valves have now operated for 10 years without any trouble. On the other hand, valves fitted to another machine had their stems hard-chromed to resist oxidation; these seized in the guide bushes as a result of severe galling of the chromium under the influence of lateral forces no greater than those with which nitrided surfaces operated satisfactorily. These instances provide an example of the occasional situation where dire predictions drawn from research work are not always reproduced in actual service operation.

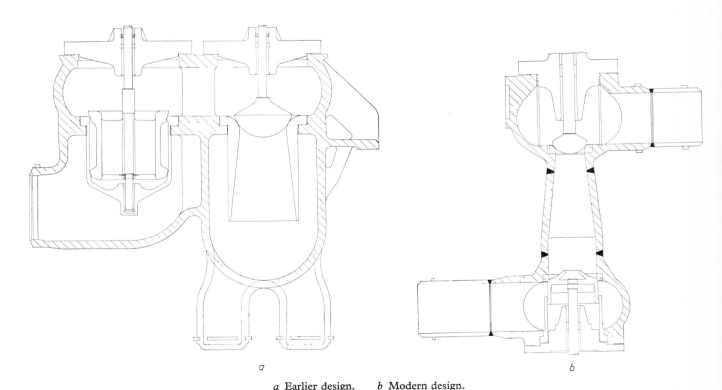

a Earlier design.   *b* Modern design.

Fig. 64.3. Comparison of casting design for reheat steam chests

Security of valve seats

Making reference again to Fig. 64.1, it will be noted that the valve seat incorporated a diffusing section and was secured to the steam chest by means of screws, which were locked against rotation by peening the adjacent metal of the valve seat hard up against their flanks. This arrangement has proved entirely satisfactory in service for valves in a number of machines up to 550-MW output, so long as the securing screws are manufactured from a creep-resistant bolt material with relatively low chromium content. In one case, however, because of the relatively poor stress-relaxation characteristics of this material at 1050°F, the screws were manufactured in a 12 per cent chromium steel. After the particular machine had been in service for just over 100 h a small number of screws worked loose in one valve seat, and the remainder fractured under their heads, apparently due to vibration of the seat and integral diffuser imposing tensile loading on the screw shanks.

It would appear that the high chromium content of the bolt steel prevented the development of rust, which provides a high degree of friction to resist any slacking off of the screw. A contributory factor may have been the thermal stressing induced not only during transient temperature conditions (which for the particular machine were at times savage) but also as a result of the small difference in coefficients of thermal expansion of the materials of the screws and valve seat. It is conceivable that this could lead to crushing under the screw heads relaxing the tension in the screw, and thus giving the seat with integral diffuser more freedom to vibrate under the excitation forces generated by the steam flow. Unfortunately, the components were too damaged to permit confirmation of the mechanism of failure.

The immediate solution was to use screws made from the material having less oxidation resistance, and to make the method for locking by peening more positive, since when there has been no further trouble.

Once again it is interesting to note that another machine of slightly larger rating and similar stop-valve temperature entered service with the original arrangement, but employing screws manufactured from the lower grade material. No trouble was encountered in the five years elapsing before the improved locking method was incorporated as a precautionary measure.

The development of alternative methods of securing valve seats has progressed very considerably since the above occurrence. Because the valve seat has to be assembled through an opening in the steam chest which is sealed off by a cover, the joint of which is subject to full stop-valve steam conditions, there is an obvious advantage in reducing the diameter of the valve seat, with the consequent benefit of reducing the diameter and sealing problems associated with the access hole. A rather more elegant construction was developed, with the seat directly welded into the steam chest. The weld was designed with flexibility to absorb differential thermal expansion, and a comprehensive series of tests showed the new design to be considerably stronger than the previous one. There was,

however, a tendency in service for the welds to develop cracks; once again, probably due to a combination of differential expansion and steam-buffeting forces.

The solution to the problem can be noted in Fig. 64.2, and it is considered to combine the best features of the two methods. The valve seat is secured by means of an annular ring screwed into a threaded recess in the steam-chest forging, and adequate provision is made for the ring to be securely locked in position. With this arrangement, the outside diameter of the seat is kept as small as possible whilst retaining the inherent ability of the screwed attachment to accommodate small differential movements. A further improvement is to machine the diffusing section integral with the valve chest, thereby eliminating possible bending moments imposed on the seat due to instability of steam flow in the diffuser.

THERMAL STRESSING

While high-temperature components of steam turbines have always been subject to thermal stressing during transient changes of temperature, the frequency of the imposition of stress cycling has been considerably increased by the general trend towards cycling or two-shifting lower merit plant, leaving higher merit plant to generate at full output. This practice was adopted in the U.K. at the end of the 1940s; as a result, U.K. manufacturers have amassed a very considerable amount of experience from machines operating under such a regime. The problem has achieved even greater significance with the adoption of higher inlet steam temperatures and pressures which necessitate the use of steels of higher creep resistance and thicker sections for pressure-containing castings.

Recently, the analysis of thermal stresses has advanced considerably, and computer programmes have been developed for standard, relatively simple geometries. This approach is of immense value in analysing the patterns of thermal stress in new designs, and the basic work is still being extended. Knowledge of the material characteristics is also being furthered by extensive programmes of laboratory testing which yield important data regarding the ability of materials to withstand thermal cycling involving elastic strain, plastic deformation, and creep.

The application of these recent analytical methods came at a time when considerable progress had been made as a result of intelligent analysis of service experience, from which the overriding lesson learnt was that components subject to thermal stressing have to be designed with sections as uniform as possible, together with uniform conditions of surface heat transfer. Thermal stressing also occurs during the casting of a component; in this respect the designer must pay equal attention to manufacturing experience in that a defect resulting from thermal stressing during cooling of the casting is likely to be extended by cycling in service.

An example of improved simplicity of casting is shown in Fig. 64.3, which compares two intermediate pressure (i.p.) turbine reheat steam chests; that shown as Fig. 64.3a

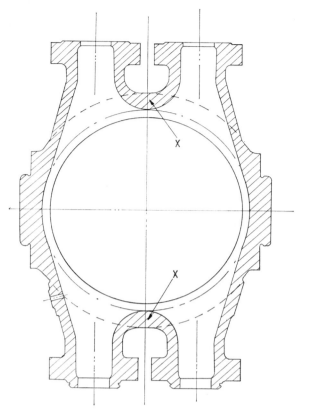

Position of cracks marked 'X'.

Fig. 64.4. Section at steam inlet belt of h.p. cylinder

being of an earlier design, and that as Fig. 64.3*b* being the modern design. In such a case it is difficult to separate the improvement in design that occurs naturally as technological advance from that fed in as a result of operational behaviour. Certainly, in the case of such steam chests, both aspects had a large part to play; the service experience acting as a spur to the development of a more uniform casting which also incorporates the benefits of improved aerodynamic performance.

Fig. 64.4 shows a cross-section through the inlet belt of the h.p. cylinder for a 60-MW non-reheat machine, having inlet steam conditions of 900 lb/in² (gauge), 900°F. The cylinder is of single case design with four steam inlets, two in each of the top and bottom halves, the material of the casting being 0·5% Mo steel. After a very large number of starting cycles (typical operation of one machine involved 2500 starts in the first 11 years of service), cracks developed in the area between adjacent inlet branches, marked 'X' on the illustration.

Thermocouples were fitted to a number of machines and the metal temperatures monitored during starts. It was found that the temperature of the metal between the inlets rose more rapidly than that in the plain cylindrical wall, probably due to a higher rate of heat transfer existing in this area. With the aim of modifying the heat transfer,

small baffles were fitted on the inside of the casting over the affected area, but the cracks continued to extend slowly with further service.

Another feature noticed from close analysis of the temperature records was that the maximum stress occurred at a temperature in the region 550–750°F. Since the preliminary evaluation of the metal before its adoption in machines had only covered its short-term elastic behaviour over the full temperature range and its long-term creep behaviour at elevated temperatures, its full properties were not fully known over the temperature range in question. Laboratory tests were therefore carried out on test specimens, and it was found that the material exhibited a considerably lower ductility in this temperature range for repeated stress cycling. Alternative materials were subjected to similar tests, and from these results an alternative composition was adopted for cylinder castings.

This provides a good instance of a new material put into service on the basis of satisfactory performance under the normal test programme. Of necessity, such a programme cannot include the full range of all possible transient conditions, and the programme omits evaluation of the material at the condition which subsequently emerges as being critical. Needless to say, as a result of such occurrences as that described above, the test programme on new materials is constantly being extended to take account of such findings.

In the U.K. the Central Electricity Generating Board (C.E.G.B.) adopts the policy of carrying out tests on one machine of each prototype design to assess the level of thermal stressing occurring during starting, and to determine the optimum mode of operation for a starting and loading cycle. In the past, it has been the practice for such tests to fit a number of thermocouples to the turbine cylinders and measure the rate of rise of temperature. From analysis of these records some indication can be obtained of the level of thermal stressing within the components. Recently, however, the approach has been extended to include the attachment of a large number of strain gauges to turbine components to measure directly the stresses that occur during starting. Such an arrangement is applied to the h.p. and i.p. cylinders of a turbine for a 500-MW set which has recently been commissioned. From careful analysis of the records, considerably greater insight will be gained into the detailed transient stress pattern which occurs when such machines are operated under a cycling regime.

TURBINE BLADING

The most common cause of failures of blades in service is fatigue resulting from excitation of a natural resonance of the blade system by periodic forces in the machine. In the last 30 years the control of blade vibration has advanced significantly, both by a careful analysis of the behaviour of blading in service and a scrutiny of any damage which may have vibration as its main or subsidiary cause; and also by careful measurements on a large number of full-scale wheels, including many running over a range of

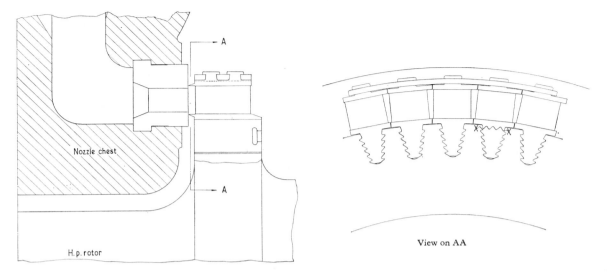

Fig. 64.5. Arrangement of h.p. first stage

speeds that includes normal service speed. The success of this approach is patent in that the number of blade failures from vibration fatigue has been considerably reduced. The techniques currently being employed by the author's company have achieved the remarkable record of only one blade failure due to this cause on those machines designed on the basis of this technology during the past 18 years.

The reason for this particular failure is of interest in that unusual temperature distribution in service caused the blade-vibration frequency to be different from that measured during tests at uniform temperature. The stage in question was in the i.p. cylinder of a turbine in a nuclear power station operating on a dual-pressure steam cycle. The introduction of steam of high temperature from the low-pressure (l.p.) source caused the mean operating temperature of the disc to be above the mean temperature of the blades. This temperature gradient caused a reduction of the blade-vibration frequencies by affecting the root fixity, and was sufficient to lower the lowest blade-vibration frequency to the region of the nozzle-impulse frequency at normal service speed.

The information provided by the investigation into this failure is, of course, of prime importance in avoiding a repetition of the circumstances in future.

First-stage h.p. blades

An area of blading which has proved to be particularly susceptible to damage in almost every manufacturer's designs is the first moving row of blades in machines provided with nozzle governing, i.e. with control valves arranged to supply separate first-stage nozzle arcs which open sequentially. With such an arrangement, operation at partial loads results in a higher level of bending stress in the moving blades, due to steam forces, than that existing at full load. There are also other disturbances to the flow

pattern as a result of the discontinuity in the steam flow around the full circumference of the inlet nozzles. Consequently, it is necessary to design the first row of moving blades in a nozzle-governed machine to be much more robust than would be the case were steam to be admitted to the full arc of nozzles over the complete load range.

A typical arrangement which has been used by the author's company is shown in Fig. 64.5. The nozzles are divided into four quadrants, with the two primary quadrants being arranged diametrically opposite, each fed through an individual control valve. The two remaining quadrants are fed sequentially through a secondary and a tertiary governor valve.

Fig. 64.6. Root of broken first-stage h.p. blade showing cracks

Fig. 64.7. Last-stage l.p. blades

The impulse blades are held in the shaft by means of an axial side-entry root of fir-tree form, which has the advantage of minimizing stress-raising features in the rim of the high-temperature section of the turbine rotor. This form of root has a very good record of service experience. In the particular instance being described, it had been established experimentally that blades were not in resonance under running conditions. Each blade is machined with an integral shroud, over the top of which is fitted a separate coverband tying the blade tips together in groups of four. After the machine had been started several times and had run for a few hundred hours at loads up to a maximum of 65 per cent c.m.r., a number of moving blades broke off at their roots. During its early life the machine ran for a short period with only one of the two primary nozzle groups in operation. Such a mode of operation materially increases the level of stress at the blade roots above the maximum design stress level, which occurs at full primary load, i.e. with steam admitted to both primary nozzle arcs.

The failures were of a fatigue nature, the majority being positioned at the neck of the blade root, indicated 'XX' in Fig. 64.5. There were also cracks in the roots of a number of other blades which had not broken through. A puzzling feature of these latter cracks was that many were not located at the theoretically most highly stressed area of the root, i.e. at the neck, but ran in a diagonal direction, as shown in Fig. 64.6.

Another feature of the early operation of the machine was the difficulty experienced in controlling the temperature of the steam from the boiler at a substantially uniform level. On a number of occasions large and rapid changes of temperature occurred, such as a fall of 240 degF in 4 min. Such an event imposes thermal stresses on the blade-root castellations and upsets the distribution of loading in the castellations in such a way as to impair the stability of the blade under the interrupted loading associated with partial admission. Moreover, despite the high degree of accuracy achieved in the machining of the blade roots and the spindle blade grooves, it is to be expected that the distribution of load between the castellations will not be perfectly uniform initially. Equal distribution is achieved as a result of primary creep during the early running life of the machine.

The successful solution adopted for this problem was to make the root section of the blades considerably stronger, to achieve a higher factor of safety than that previously considered necessary. Furthermore, commissioning procedures were modified to run any machine designed for nozzle cut-out governing under conditions of throttle-governing control for the first few hundred hours of its life. This was to permit the achievement of uniform distribution of the load on the castellations of the fir-tree root to be achieved by the mechanism of primary creep.

Last row l.p. blades

There is no need to emphasize the necessity for very careful analysis of each design of the very long highly loaded blades required for the last l.p. stage of large machines. Although much can be done to predict the vibration characteristics, it is the practice of most, if not all, turbine manufacturers to build prototype wheels for dynamic vibration testing. This ensures that no important resonant frequency occurs at a low multiple of running speed frequency. There is also a general move to avoid methods of vibration control which introduce unwanted stress-raisers in the blading, such as, for example, holes for lacing wires. The arrangement shown in Fig. 64.7, adopted by the author's company, employs an elastic arch coverband, which gives excellent control of vibration. Moreover, since it is located close to the tip of the blade, in the region of low centrifugal stress, the deleterious effects of stress concentration of the small holes for the rivets used to attach the coverband pieces to the blades are of no serious consequence.

With the continued growth in machine rating, the increasing shortage of sites offering an unlimited supply of cooling water to the condenser, and the cost of building the largest condensers to achieve the maximum vacuum obtainable, it often becomes economically attractive to design power stations that have the condenser back pressure considerably greater than has been the case in the past.

The resulting larger mass flow per unit area of the last row blade annulus significantly increases the steam loading and, therefore, the bending stress in the last row blades. The steam loading gives a random mixture of forced excitation, and it is important to know the level of vibration randomly excited in this way in bladed wheels in service.

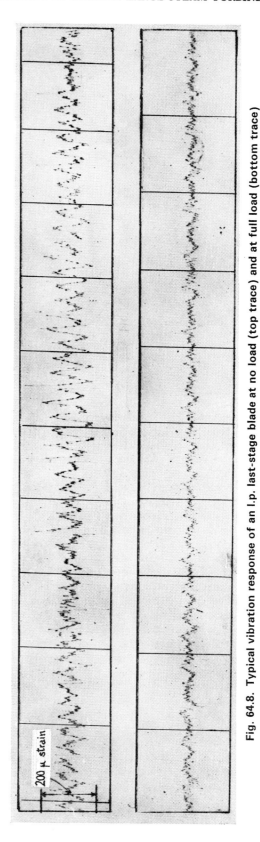

Fig. 64.8. Typical vibration response of an l.p. last-stage blade at no load (top trace) and at full load (bottom trace)

200 μ strain

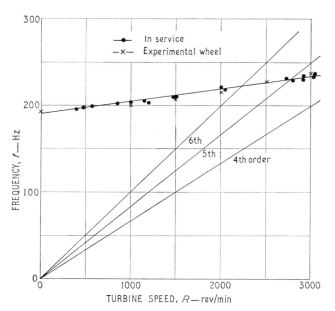

Fig. 64.9. Comparison of fundamental blade-vibration frequencies measured on l.p. blades in service and on an experimental wheel on test

In the U.K. the C.E.G.B. has co-operated in making available machines that permit strain gauges to be fixed to the last row blades. From these, leads may be taken either to slip-rings (3) located near the end of the shaft, or to radio transmitters (4) attached to the last row bladed disc or to the nearest coupling, which pass signals to a neighbouring stationary pick-up ring.

The author's company has used both arrangements of the radio telemetering system, and a typical example of a stress record obtained is shown in Fig. 64.8 (5). The trace shows strain amplitude against time, the scale of 200 microstrain being shown on the figure, and the total time elapsed on each trace being about 0·2 s. This particular record demonstrates the randomly excited fundamental blade vibration. The information it contains is of great use in guiding the mechanical design of blades to withstand steam buffeting, and in establishing the steam load that long blades can carry. These service tests have demonstrated that well-designed blading operates in service with very low values of alternating stress.

Another use of the information obtained from machines running in service is to check the validity of tests made on prototype stages in laboratory test plant. This has been done for stages in service using highly sensitive crystal strain gauges, capable of detecting the very low levels of self-excited vibration, connected to radio transmitters. Fig. 64.9 shows a typical result obtained from a 350-MW last-stage wheel in service and from the corresponding test wheel in the laboratory. The agreement is striking.

EROSION OF LOW-PRESSURE BLADES
The problems associated with the erosion of the tips of

high-speed l.p. blades due to the impingement of water droplets entrained in the steam have been described in a number of other papers (6)–(9). A very considerable amount of research work has been carried out in laboratories using specially constructed test rigs. This work has produced valuable data regarding the mechanism of erosion and qualitative comparison between alternative materials for blade protection. The difficulty exists, however, that it is impossible to simulate exactly in a laboratory rig the flow conditions, droplet size, distribution, etc., that exist in a real machine. Moreover, this could not be readily reproduced in a model l.p. turbine, since initial droplet sizes and the subsequent water-deposition rates on to the blading are influenced by the rate of expansion. In the case of a one-third scale model, this would be three times that in a full-scale machine. It is therefore necessary to make observations and take measurements on full-scale machines in service operation.

Service experience on recent large turbines, in which design to combat erosion has had to be based on largely empirical assessment of results from earlier machines, has been very encouraging. Fig. 64.10 shows the blading of a 500-MW unit after 11 300 h service, with nominal exhaust wetness of 9 per cent and peripheral speed of 1780 ft/s.

In recent years, understanding of the basic mechanisms involved has been enhanced very considerably by the use of introscopes positioned downstream of the last moving row of blades. By this means, observations can be made of the pattern of water flow on the last stage of fixed blading. It has been possible to compare the change in behaviour resulting from various modifications to the geometry, both of the blade path and the actual blade profiles. This work is currently being extended, and it can be expected to yield more fundamental information of direct application to the design of yet larger turbines employing last row blades with higher tip speeds made possible by the continued development of stronger blade material.

CONCLUSION
In this paper it has only been possible to discuss a few examples showing how service experience has contributed to the development of larger and more reliable steam turbines. Furthermore, these have been drawn solely from the experience accumulated by the author's company. It is hoped that this paper will encourage the presentation of further examples drawn from the experience of other manufacturers.

Failures of components in machines in service often result in long and very expensive outages. In addition, the loss of high merit generating capacity of a modern large unit can amount to £20,000 per day, solely as a result of providing the system demand from less efficient stations. It is therefore essential that experience gained from service difficulties be carefully analysed and fed into the design activity, so that maximum advantage is taken of potentially expensive lessons. Measurements taken on machines in service must receive equally detailed analysis to obtain

Fig. 64.10. Extent of erosion on l.p. blades after 11 300 h service

full advantage in incorporating the data in the design of new machines.

ACKNOWLEDGEMENTS

The author wishes to express his thanks to C. A. Parsons & Company Limited for permission to publish the paper, and to many colleagues for the help they have given in its preparation.

APPENDIX 64.1

REFERENCES

(1) STRAUB, F. G. 'Blue blush characteristics', *Proc. American Power Conf.* 1955 **17**, 511.

(2) PETERSON, R. E. *Stress concentration design factors* 1953 (John Wiley & Sons Inc., New York).

(3) WIESE, P. R. and DAVID, T. J. 'Collecting data off high-speed rotors', *Engineering, Lond.* 1963 (27th December).

(4) JONES, D. H. 'Multichannel contactless telemetering system for vibration studies on steam turbine blades', *Int. Telemetering Conf.* 1963 (September).

(5) CAVE, L. E., NORMAN, J. R. and LUCK, G. A. 'Vibration measurements on some stages of a 120-MW steam turbine', *Engineer, Lond.* 1964 (25th September).

(6) GARDNER, F. W. 'The erosion of steam turbine blades', *Engineer, Lond.* 1932 (5th–7th February).

(7) SMITH, A. 'Physical aspects of blade erosion by wet steam in turbines', *Phil. Trans. R. Soc., Series A* 1966 **260**, 209.

(8) SMITH, A., KENT, R. P. and ARMSTRONG, R. L. 'Erosion of steam turbine blade-shield materials', A.S.T.M. Special Tech. Publ. No. 408, 1966.

(9) ELLIOTT, D. E., MARRIOTT, J. B. and SMITH, A. 'Comparison of erosion resistance of standard steam turbine blade and shield materials on four test rigs', A.S.T.M. Special Tech. Publ. No. 474, 1969.

TAKING ADVANTAGE OF EXPERIENCE GAINED IN THE APPLICATION OF AUTOMATIC CONTROL TO MARINE MACHINERY

R. L. DENNETT*

The experience gained during dock trials, sea trials, and subsequent operation in service of complex, modern, controlled machinery installations clearly demonstrates the need for careful planning and co-ordination of effort if costly delays, coupled with unsatisfactory performance under operating conditions, are to be avoided. In most cases the design concepts are good, but as a result of failure to communicate, say, environmental requirements, many control systems fail to produce the results expected. On many occasions the individual problems are only apparent at a late stage. Intensive effort is then applied by all concerned. At best, a compromise solution is found that will enable the installation to operate, although not quite in the manner originally envisaged.

This paper attempts to locate areas where the saving of misplaced effort might be achieved, to emphasize the necessity for a complete understanding of the overall plant requirement by all parties concerned, and to suggest how close co-operation to attain these ends might be maintained. To emphasize some of the points, actual experiences are quoted, showing how apparently minor omissions or divergence from basic requirements can have widespread effect.

INTRODUCTION

DURING THE DEVELOPMENT in recent years of more advanced automatic control layouts in the field of marine engineering, operation, or maloperation, as the case may be, does not appear to have attributed to any major catastrophe. Thus, fortunately, it has not been possible to obtain pictorial evidence of what may occur if satisfactory precautions are not taken at all stages of development. This is not to say, however, that problems encountered have only been of a minor nature. Considerable embarrassment, irritation, and frustration, not to mention extra work and cost, have been experienced in, firstly, establishing and then eliminating defects which, in retrospect, should never have been encountered.

The British Ship Research Association (B.S.R.A.) in their *Recommended code of procedure for marine instrumentation and control equipment* have produced a guide which, if their advice is followed, must benefit the industry as a whole. In addition, detailed and interesting papers covering all aspects of automatic control system philosophy, design, and operation have been read from time to time. It is not the intention here to try to emulate these authors.

The MS. of this paper was received at the Institution on 20th October 1970 and accepted for publication on 25th November 1970. 34
* *Assistant Manager, Test and Service Department, Foster Wheeler John Brown Boilers Ltd, Greater London House, Hampstead Road, London N.W.1.*

However, in the following paragraphs some personal experiences are quoted to illustrate the types of problems which could have been minimized, if not eliminated, by adherence to principles discussed in those publications.

In general, the phases of development are specification, design, installation, commissioning, and operation—with co-ordination firmly controlling the whole project. These phases are dealt with separately.

SPECIFICATION

Before any control system is designed there must be laid down an exact plan of what it is required to achieve, and the parameters within which it is to operate. Until this is done there will always remain the possibility that future problems may be resolved locally but out of context with, and even to the detriment of, the overall scheme.

A comprehensive logic diagram will incorporate all functions of the components of a system in their correctly required sequence of operation. Reference to this, and observance of its requirements at all times, will ensure that no modification will be made to a component if it is not relevant to the scheme.

Many times in the past it has happened that modifications to components have been made, or instruments added or maybe removed, without the reasons for modification being adequately recorded. Eventually, a time comes when a query is raised on the necessity of such an item. It is then found difficult or, at best, time consuming to ascertain why it was changed; or, in fact, from what it was changed.

It is essential that all authorized modifications are transmitted to, and incorporated in, copies of relevant drawings issued to sub-contractors.

CO-ORDINATION

The responsibility for the whole of a control installation should lie with one co-ordinating authority.

The ultimate responsibility, of course, will remain with the shipbuilder. However, in this age of consortia, merger, and rationalization not every shipbuilder may have the facilities necessary to carry through a project in every detail, and he may consider it expedient to contract out, in varying degree, to an outside specialist. This method can be of considerable benefit in avoiding much duplication of effort, and it should ensure that an accumulating quantity of data and experience are available to the whole industry.

All information covering the system should flow to or from the selected authority and not directly between suppliers of equipment.

It is all too easy to lose essential information by trying to take short cuts, as may be seen from the following examples.

Example 1

In this case, a high-level, priority meeting was convened to inquire into the problems experienced in commissioning the combustion control system installed in the subject vessel. It was apparent that each sub-contractor had rigorously checked his design characteristics against original requirements and confirmed that all parts supplied were correct to drawings. However, when the original requirements were tabled it was evident that the burner curve supplied to the main contractor differed from that supplied to the control manufacturer. One of the curves had a square root characteristic whilst the other was linear.

Although simple in origin, the correction of this defect resulted in lost time during commissioning whilst modifications were carried out. Had there been a central co-ordinating authority with definite responsibility, this delay would not have occurred. The matching of burner curves issued would have been checked and confirmed before orders were placed.

Example 2

With rationalization of design and the reduction of design margins, extreme care must be taken to ensure that all component characteristics and requirements are known and incorporated correctly into the scheme.

In this case, the minimum and maximum steam requirements of a system were calculated. A burner turn-down range was determined which allowed for all burners to be in use throughout the control range. When sea trials were carried out, however, it was established that the minimum steam demand was lower than that calculated, and it was necessary to shut off burners to achieve the required operating range. The combustion control system had to be modified to incorporate sequencing of burners in order to accommodate the extended control range. All following ships in the series, in various stages of completion, were similarly modified.

It is debatable whether this defect could have been found earlier, as it was the result of an error in calculations based on the information then available. Nevertheless, it is considered a type of problem which should be eradicated as experience is built up and a closer interchange of knowledge is developed.

DESIGN
Compatibility of equipment

In a well-designed installation the machinery being controlled and the control system itself must be complementary to each other. No amount of sophisticated control equipment can improve the operation of a machine beyond its built-in capability. No machine can give optimum performance over a range of load without good control.

Until a future time when complete reliability of machinery and controls has been achieved there must always be compromise, and this will invariably entail duplication of some sort. Duplication of components will inevitably complicate a system with an agglomeration of change-over valves which may be manually operated, and therefore subject to possible misuse. Duplication of equipment with manually operated stations involves the utilization of personnel, which negates to some extent the advantages of providing installations.

This, of course, is within the province of the manpower requirements necessary to run and maintain a system employing automatic control; a subject obviously taken into account in the design philosophy. What is not readily assessable is the time required for operating engineers to become fully conversant with the system, and to be able to diagnose correctly and put right a fault in any component before the system as a whole is adversely affected.

Paradoxically, while increased reliability is being built into machinery the opportunity for operators to learn from experiencing faults is diminishing. Therefore, in the event of a fault occurring they have to rely on knowledge acquired from instruction books (which of necessity must be of a high standard), coupled with, where possible, technical courses at manufacturers' works. Any such training can never have the impact of personal experience. It may be argued that immediate corrective action by an operator is often only possible after a fault has been experienced by him for the second time. This may be once too often.

Environment

An important point that should not be overlooked is the physical positioning of control equipment in machinery spaces. Components which in emergency may have to be manually operated must be fitted in accessible places. In any event, accessibility must be provided for future maintenance. No instrument should be fitted in a position where temperature and humidity, for example, could exceed that which is acceptable to the instrument. This may seem a very elementary precaution, but it is one that often appears to be considered secondary to its location with respect to pipe systems and machinery layout. If it is necessary, owing to other factors, to position instruments in remote places, adequate ventilation should be arranged. Failure to do this may lead to erratic operation or complete breakdown, with the extent of the after-effects dependent on the function of the item in the system.

First principles

It is essential that control system designers should not neglect engineering first principles. Many older engineers have learned this the hard way; but it is frequently taken for granted, or even overlooked, by more modern students of engineering, amid the ever-increasing mass of technical knowledge which they must attempt to assimilate.

Time and again the importance of the absolute boiler water-level gauge glass has been highlighted by accidents due to too much reliance being placed on remote reading instruments. And this in spite of the fact that bitter experience in the past led to the introduction of routines such as 'gauge glass drill' to ensure that the absolute glasses were reading correctly.

From a boiler designer's point of view the control room watch-keeper should have an unobstructed view of the absolute glasses to enable him to check immediately the reading of his remote panel gauges and the operation of his water-level controls. These instruments must be independent of each other to avoid a false reading initiating a false control signal.

Example

Quite recently a modern single-boilered oil tanker was out of service for an extended period to replace a large number of boiler tubes following damage due to loss of water.

The trouble was twofold. The two-element feed controller (Fig. 70.1) utilized a differential-pressure instrument to measure variations in boiler water level against a constant head of condensate in a reference leg. Leakage of the drain valve at the lower end of the reference leg resulted in a reducing level in this leg. The reducing differential pressure caused the feed control valve to shut in, owing to an apparent high level in the drum. In fact, the slowly dropping level in the reference leg led to a corresponding drop in the drum level. Water-level indication on the control panel and low-level alarm were provided by the same signal, so these showed normal conditions.

A separate low-level oil shut-off system had been provided, and this was operated by a float–magnet switch. This failed to operate, the float being subsequently found

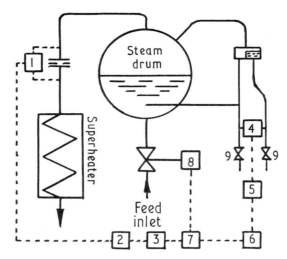

1 Steam flow transmitter.
2 Square root extractor.
3 Biasing relay.
4 Level transmitter.
5 Derivative unit.
6 Recorder/controller.
7 Computing relay.
8 Feed control valve.
9 Drain valves.

Fig. 70.1. Two-element feed control system

to have become restricted in movement by sludge and metallic particles.

Had it not been possible for the control room watch-keeper to verify quickly the low boiler-water level by means of the absolute gauge glasses, after he had noted the gradual reduction of recorded feed flow, the damage to the unit, although already serious, would have been far more severe.

INSTALLATION

It is necessary to convey to personnel at all levels of the manufacturing structure the importance of details in design which may to some appear superfluous but which have been incorporated for very valid reasons. These reasons may apply not only to the component itself but also to effects on other parts of the overall plant.

Example

A case in point is the pressure tightness of boiler gas casings; and one might ask how the pressure tightness of boiler gas casings can affect the operation of an automatic control system.

In the past, balanced-draught installations carried gas space pressures below atmospheric; thus, any leakage was of air from without to within the boiler casings. However, with the swing to forced-draught installations, necessitating positively pressurized gas spaces, the outward leakage of gas and soot had to be restrained by providing air-pressurized double casings. Nevertheless, air leakage into the gas spaces from these pressurized areas was still acceptable within limits.

With the advent of higher combustion efficiencies, such leakage of air into the combustion chamber had to be eliminated. Any ingress of air downstream of the burner registers will have a diluting effect on the combustion gases, and thus affect the gas analysis. As the settings of the automatic combustion controls are adjusted in accordance with this gas analysis, the actual air-to-fuel ratio at the burners will be too low and flame shape will not be as anticipated. In fact, a situation may be reached where combustion is not completed without utilizing the inward leaking air.

This problem was recently encountered on a series of vessels where 6 per cent excess air in flue gas was specified. On detailed examination it was found that the sealing plates around water wall headers had not been fitted according to design. As there were eight headers on each boiler, the quantity of air leakage resulting was considerable.

The fact that such defects are eventually hidden behind the outer casings is no excuse for their presence, and is a reflection on the efficiency both of labour and of quality control during the construction stage.

COMMISSIONING

Once the design philosophy of a plant has been established and the various components have been selected and matched to one another, the key to the problem of satisfactory operation in service lies in the commissioning and trials procedure. If the responsibility of design co-ordination is subcontracted to a specialist company, it may well be of advantage to arrange for this company to be responsible for the commissioning and trials procedure, supplying a commissioning team, as opposed simply to an adviser. This will place the responsibility for the project in one place and avoid endless argument between suppliers of different components.

A complete functional check-out of integrated control systems prior to sea trials should be the aim. From the outset, the time allocated for these checks should be considered inviolate. Too often this appears to be a useful contingency period in which to absorb earlier delays in other fields. The fully integrated control system cannot be considered the poor relation of marine engineering, and the idea that 'if everything else works the controls can be adjusted later' is not realistic.

Example

The contamination of the feed systems, the boilers, and parts of the steam and drain systems of a vessel during trials might have been prevented if sufficient time had been allocated to proving components and alarms in the system (Fig. 70.2).

The trouble stemmed from the fact that purge steam non-return valves on the oil fuel burners were not pressure tight. Under normal operating conditions this resulted in oil leaking back into the purge steam manifold. From this point, two paths were open. One was back to the low-pressure steam range through manual shut-off valves

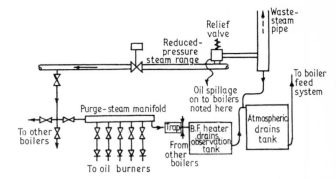

Fig. 70.2. Burner purge steam system (original)

which had been left open. The first sign in the boiler room that there was anything amiss was fuel oil leaking from the steam-line relief valve and on to the boiler roof panels. By this time oil was noticed in the atmospheric drains tank, and the job was shut right in while investigations were carried out.

The second path was then discovered. At the end of the purge steam manifold a steam trap was fitted to clear any condensate. The condensate line was led to the heater drain observation tank at the side of the boiler room, and from thence to the atmospheric drains tank in the engine room. Oil from the manifold had obviously followed this path. An alarm at the observation tank had not been commissioned, and the tank itself was not in a readily visible position, so contamination at this point was not noted immediately.

Apart from the fact that the purge steam non-return valves and the observation tank alarm system had not been checked out, two other aspects were highlighted and remedied. These were the fitting of drain cocks with which to prove the drain system clear; and, more important, the modification of the purge steam system to ensure that the steam originated from the high-pressure steam range, where the pressure would always be higher than the maximum obtainable oil pressure (Fig. 70.3).

OPERATION

Successful operation in service is dependent on good maintenance of equipment which has initially been proven

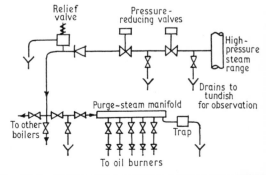

Fig. 70.3. Burner purge steam system (modified)

correct. This is borne out in the following case. Other factors covered in previous paragraphs are also relevant here.

A large oil tanker was recently lying at anchor, fully loaded, suffering from a spate of steam-joint leakage and superheater tube failure. A condition had been reached where it was necessary to shut down one boiler or the other every few days for repairs. In addition, the ship's engineers had worked themselves almost to a standstill.

A shore squad was brought on board, as many joints as practicable were renewed, and all suspect superheater elements were plugged. An investigation was carried out to establish the reason for failure, the results of which follow:

(a) The condition of the furnace showed that combustion conditions had been extremely bad and heavy flame impingement had occurred in the area of the damaged superheater tubes. The burner quarls were badly burned and flame shape could never have been acceptable.

(b) An inspection of the registers, which were roof-mounted, showed that due to failure of air seals no pneumatic actuators were operable, electrical limit switches had literally disintegrated, and in fact several air slides were mechanically seized in the open position.

(c) It was reported that the combustion control equipment did not regulate the air supply, and an examination showed that the furnace pressure tapping points were choked with carbon.

(d) The method of sequential burner operation during manoeuvring was for a man on receipt of a signal, which was in fact a blow of a hammer on a girder, to close the oil-supply valve to a particular burner, the air slide being left open all the time.

Condition (a), which was responsible for the continued superheater tube failure, was directly attributable to the combined effects of (b), (c), and (d).

Defects noted in (b) were directly attributable to the inability of components to withstand the conditions of their environment.

The defect noted in (c) should have been self-evident, but the ship's engineers had made an incorrect diagnosis which led them to suspect the instrument settings. These had been readjusted and were found hopelessly out of balance.

Defects (b) and (c), not having been corrected, led to the adoption of the routine mentioned in (d).

The continued leakage of steam joints had been an aggravating feature of this case in that the available manpower became over-extended and could not give adequate attention to routine maintenance. The solution here was to provide jointing material of a higher standard.

Corrective action was taken and the vessel has since steamed satisfactorily, although three additional engineers were embarked to assist with other running repairs.

CONCLUSION

The conclusion to be drawn from the various experiences lies in fact in the Introduction to the B.S.R.A. Code, paragraph 4 of which states, 'Instrumentation and control schemes must be considered as an integral part of marine systems, no less important than the machinery to be controlled and monitored. Meticulous attention to design, installation, testing, and commissioning will produce reliable systems, and reduce installation and maintenance costs. All this work demands trained personnel, and consideration should be given to the provision of facilities for educating and training instrumentation and control engineers for the marine industries.'

These industries cannot afford to ignore that advice if the above experiences are not to be repeated.

ACKNOWLEDGEMENTS

Acknowledgements are due to the Directors of Foster Wheeler John Brown Boilers Limited for permission to publish this paper, and for the assistance given by the Company in its preparation. Thanks are also due to B.S.R.A. for permission to quote from their publication, and to builders and operators of ships from which the experiences were gained.

C71/71 SPECIAL PROCEDURES FOR ACCEPTING NEW WARSHIPS INTO SERVICE

R. M. INCHES*

During the last 15 years, warship propulsion and associated systems have become not only technically more advanced and complicated but also more tightly integrated. In addition, automatic control of systems has been widely introduced. Commissioning new ships, and even refitted ships, has in consequence become much more difficult. The paper sets out the new approach to commissioning which the Royal Navy has adopted to deal with this situation. The new scheme is comprehensive and controlled from a headquarters. It includes a set programme, specialist checking teams, elaborate advance operator training, and an early assessment of the technical–logistic requirement.

INTRODUCTION

UNTIL ABOUT 10 years ago the acceptance into service of the propulsion and associated systems of a warship was a comparatively straightforward process, involving primarily the contractor, the ship's company designate, and an experienced officer on the staff of the accepting authority. As long as these systems in successive classes of ships continued to be broadly of the same kind, and closely related in design, this approach worked quite well. However, the diversification in the post-war designs, such as *Weapon*- and *Daring*-class destroyers, followed by the moderate change in philosophy and the great change in engineering design in the *Whitby*-class frigates, produced evidence that the approach was under considerable strain. The next development was the introduction of gas turbines into propulsion, in the general-purpose frigates and guided-missile destroyers, paralleled by the increasing importance of ancillary systems, such as air conditioning, stabilizers, and aircraft fuel. Superimposed on all this were automatic controls. Clearly, this could not be dealt with on the old basis.

A completely new and comprehensive scheme for working new ships up to acceptance level, and proving that they met the standards required, was therefore instituted. It consists of four main elements:

(1) A special commissioning trials programme.

(2) A machinery trials unit (M.T.U.), with specially experienced officers and ratings responsible for checking the performance of ships under the trials programme.

(3) Elaborate simulators in training establishments, set up well before the machinery simulated goes to sea, to provide practical training, first for the M.T.U. staff and then for operators designate.

(4) Advance assessment of the logistic load.

SPECIAL COMMISSIONING TRIALS PROGRAMME

The commissioning programme of an R.N. ship is divided into four parts, which cover the following periods:

Part I: Launch to contractor's sea trials (C.S.T.)
Part II: Contractor's sea trials.
Part III: Contractor's sea trials to acceptance.
Part IV: Acceptance to date of operational service.

The preparation and reproduction of the first three parts of this programme are contractual responsibilities of the shipbuilder. Part I is submitted as soon as possible after launch, and Parts II and III at least three months before Part I is due to finish. The preparation of the Part IV programme is the responsibility of the senior officer of the ship. Parts I to IV are each submitted to and approved by MOD(N).

The marine engineering trials are undertaken at stages throughout the first three parts of the programme and culminate with an 'acceptance demonstration' of the machinery, just prior to the acceptance date of the ship.

Besides the various trials undertaken, three 'machinery installation inspections' are made. The object of these inspections is as follows:

(*a*) To ensure that the machinery is installed in accordance with the relevant drawings and specifications.

(*b*) To check accessibility for maintenance of components.

(*c*) To ensure that the machinery is ready in all respects for subsequent trials.

The MS. of this paper was received at the Institution on 20th October 1970 and accepted for publication on 25th November 1970. 33
* *Captain, Royal Navy. Superintendent, Admiralty Marine Engineering Establishment, Gosport, Hants. PO12 2AF.*

The normal sequence of the machinery trials and inspections, which are conducted by the M.T.U., for a typical R.N. frigate is shown in Appendix 71.1. For the controls part of the trials the M.T.U. has available the special knowledge and experience of the machinery controls trials team (M.C.T.T.). More will be said about this team later in the paper.

During C.S.Ts the machinery trials have precedence, but the opportunity is taken to undertake other necessary sea trials such as prewetting, E.M. log calibrations, habitability, stabilization, etc., to prove the ancillary systems and equipments.

MACHINERY TRIALS UNIT

The setting up of an M.T.U. and the provision of specially trained staff for carrying out the trials already mentioned were implemented on the recommendation of an R.N. committee. This committee had been formed to investigate the kind of problem brought out in the first part of this paper.

The terms of reference of the M.T.U. are as follows:

(a) To supervise trials of machinery in: (i) new construction surface ships (for R.N., Commonwealth, and foreign navies, and for fleet auxiliaries); (ii) existing ships when new design machinery is installed, or when new performance figures are required; (iii) modernized, converted, or long refitted surface ships.

(b) To carry out machinery installation inspections.

(c) To attend terminal date and final inspections of ships to advise on marine engineering aspects in the light of inspections and trials results.

(d) To nominate the particular items of machinery to be opened up for examination after C.S.Ts.

(e) To witness a demonstration of the main machinery for acceptance (when the ship transfers from shipbuilder to customer).

(f) To render reports on the results of the trials and inspections (most trials have special forms of documentation of all important readings taken).

(g) To write trials requirements (these are forwarded to the shipbuilder well in advance of any trial).

To meet the above commitments there is an establishment of two trials teams, each usually consisting of a trials officer (commander, R.N.); an assistant trials officer (lieutenant, R.N.); a records officer (retired R.N. officer); a senior technician of technical grade II; and a chief marine engineering artificer (propulsion) (a senior technical rating).

The deployment of the trials team in the ship is basically: trials officer in the machinery control room conducting trial; assistant trials officer in the boiler room; records officer in a convenient office collating the readings taken by the recording staff (personnel from the shipbuilder, or dockyard drawing office staff); technical grade II technician as directed for a particular trial by the trials officer, and who also collates list of defects; chief marine engineering artificer checking, and undertaking, trials of auxiliary machinery. However, this will of course vary with different propulsion systems.

If the ship has been built in a Royal Dockyard the ship is steamed by R.N. personnel; but if built in a shipyard it is manned by the shipbuilder. In each case the trials officer from the M.T.U. is a serving R.N. officer, normally having the rank of commander.

The aim of the M.T.U. is the minimizing of hazards and delays in bringing ships forward, whilst maintaining the highest possible standard of acceptance of machinery for fleet service.

The advantages to be gained by having a centralized trials unit undertake all the machinery trials are:

(a) The personnel can become more experienced and specialized in the conduct of trials than was the case when the undertaking of trials was left to persons for whom trials were a small section of their work, carried out only relatively infrequently.

(b) By comparison of the standards achieved by various shipbuilders a centralized body can raise the acceptance standard to the highest practicable level.

(c) An outside authority carries more weight and can exert more pressure than those who work locally and are bound to be involved in the shipbuilder's life and problems.

(d) If there was no M.T.U. the trials would be undertaken by personnel having a 'production' rather than a 'user' background.

(e) A central organization can best keep 'trials requirements' under constant review and is probably the best authority to take the lead in writing these requirements for new design ships and machinery.

(f) By 'bowling out' the notorious snags before they even arise the process of bringing a ship forward can be speeded up. The M.T.U. is well placed to do this.

(g) If the reputation of the acceptance team is high, this gives confidence to those who have to take over the ship and run it in the future; the M.T.U. is favourably placed to establish a good reputation.

(h) The M.T.U. can keep under constant, centralized review all check-off lists and other installation aids.

(j) The M.T.U. is in constant touch with those responsible for design, and a two-way-feed system is worked which minimizes the delays in transmitting lessons learnt and improvements proposed respectively.

As an extension of the M.T.U. concept, for remote and automatic controls there is an additional team of specialists. This is the M.C.T.T., whose terms of reference are:

(a) To give guidance to contractors in installing, setting to work, and tuning control systems up to an acceptable level of performance.

(b) To give advice to accepting authorities on the level of performance achieved and where necessary acting as the acceptors.

(c) To assist in the solution of controls problems in ships in service.

(d) To feed back to the ship department lessons learnt

from the above activities and to contribute to the embodiment of these lessons in specifications and designs for future systems.

This team is part of the specialist controls section in the ship department and thus is very well placed to make the contribution at (d) above. Since the field is not only new but also changing very rapidly, this 'closing the loop' is most important.

The responsibility for the conduct of the trials still remains firmly with the main machinery contractor shipbuilder or administrative or refitting authority. In certain instances the M.C.T.T. assist in the taking of records.

The M.T.U. and M.C.T.T. are both stationed at Admiralty Marine Engineering Establishment (A.M.E.E.), but are MOD(N) outstations under the policy directive of the Directorate of Warship Design and Director of Engineering respectively. Library record facilities and administrative and technical support are available from the host establishment. This arrangement has many advantages. The most important are that it facilitates the use of the resources of the marine engineering school at H.M.S. *Sultan* to train field personnel of the trials teams and at the same time provides an invaluable feedback of experience from newly constructed ships into the research, development, and testing work in the A.M.E.E. Occasionally, A.M.E.E. personnel can be used to fill gaps in teams in a manner better than could outsiders.

TRAINING THE OPERATOR

H.M.S. *Sultan*, the R.N. principal mechanical engineering training establishment, rapidly developed machinery demonstration displays in which complete propulsion installations could be operated under advanced steam conditions of pressure and temperatures when these came into use in the fleet. The components were not necessarily those to be met with in the latest classes of ships, and there was of course an environmental limitation. Nevertheless, the prototype and early production models of most equipments were available for static demonstration and maintenance. In the later stages of the training programme there can be complete running installations of propulsion and electric generator rigs.

Though it is by no means easy to ensure that these training facilities are set up far enough in advance of new seagoing equipments, the additional pressure from the M.T.U., who must be ahead of all other users, is powerful support for the training side in their battle to achieve this.

A very sophisticated machinery control room simulator for a D.L.G. and a very complete controls laboratory have been built up to consolidate the vast amount of software which the M.T.U., M.C.T.T., and H.M.S. *Sultan* have produced.

Apart from the basic marine engineering course which deals with the general principles of the modern machinery installations, each operator and maintainer is given, wherever practicable, a pre-commissioning course before joining a type of ship in which he has not previously served, even when the ship design as such has long been in service. Since the facilities for this are there anyway, from the time when the design was new, this involves no special capital expenditure, though the running costs, and the cost of the trainees' time, are not inconsiderable. The return for this investment is not easy to assess, since the catastrophic failures, delays, and insufficiencies which these courses have prevented are not an easily recoverable statistic. However, it is probably not an exaggeration to say that without these courses the availability factor of fleet ships would have been drastically lowered and support facilities overwhelmed with breakdown repairs.

ADVANCE ASSESSMENT OF LOGISTIC LOAD

Other facets of the preparatory work for accepting new equipment into the fleet are those of logistic planning and maintenance evaluation. Each equipment is given a maintenance envelope—literally, a volume of machinery space necessary for all the maintenance functions, including repairs by replacement, to take place. (No longer should it be necessary to strip down adjacent machinery to permit equipment repair, maintenance, or service!)

Using the maintenance envelope, an assessment is made of the quality and quantity of manpower, special tools, and time required for those maintenance tasks which it is envisaged will be carried out *in situ*. A forecast is also made of the likely frequency of these tasks, allowing a ship logistic load forecast for the equipment to be built up. Spare gear requirements are added to this to complete the sea-going logistic support requirement picture. The whole is subject to review in the light of experience, through the Ship Maintenance Authority (S.M.A.).

This task is normally carried out on the prototype unit sent to A.M.E.E. for long-term evaluation, and the work is shared by A.M.E.E. and S.M.A. However, there is close contact with the manufacturer. In certain cases where extensive trials at the maker's works are required anyway, the maintenance evaluation is carried out there by a special team set up as an 'outpost' of the S.M.A. For instance, that arrangement is being used for the type 42 main propulsion machinery at R.-R., Anstey.

Additional measures for the future that can already be seen to be needed concern electronic controls and fluidics.

Electronic controls

The type 42 frigate will have electronic controls developed from Hawker Siddeley Dynamics Ltd concepts to suit the particular demands of R.N. operation and maintenance, instead of the pneumatic control systems fitted in the older ships. Essentially, the control system is built up of 'modules', the fault diagnostic techniques down to 'module' level being within the capacity of a marine engineering artificer (propulsion) [M.E.A.(P.)], who has been given a short (12-week) machinery and controls course. The modules are made up from electronic cards. Whilst the module can be changed for a serviceable one by an M.E.A.(P.), further diagnostic repair and maintenance work within the module will be the task of an ordinance electrical artificer (O.E.A.), who has been given a special course in machinery operation and controls in addition to his extensive electronic knowledge.

It will be seen that as the training of the M.E.A.(P.) and the O.E.A. is being shared by H.M.S. *Collingwood* (the R.N. electrical school) and H.M.S. *Sultan*, we are therefore moving toward a requirement for a 'systems engineer' within the ship who will have in his charge both the mechanical and electronic aspects of the propulsion system. Currently, the marine engineer will remain the heavy engineering expert, and the weapon electrical officer the electronic expert. However, it seems reasonable to forecast that this shared responsibility at management level will not be the ultimate solution to the problem.

Fluidics

Fluidics is less far removed from the normal field of the propulsion engineer than electronics, but nevertheless it has a number of novel aspects. Therefore, in advance of its introduction into the fleet, a fluidics section has been set up in the controls laboratory of H.M.S. *Sultan*. This consists of working models demonstrating the principles of fluidics, and including sensing, logic, and amplification devices. Methods of linking up with other control techniques are also shown. For the time being, only systems using air as working fluid are used, because that is what the fleet will have, but within this field H.M.S. *Sultan* is in a leading position.

CONCLUSIONS

The trend towards integrated system designs and 'packaged' equipment groups has, if anything, underlined the need for preparatory work and special arrangements of the kind under discussion. Not only are the repercussions of mishap or mishandling more extensive but also the tighter integration complicates the task of familiarization. It can therefore be predicted with confidence that the arrangements the R.N. has made will continue, though their form will no doubt change somewhat to meet the changes in the fleet. Currently, and for the next few years, the load is heavy, and any suggestion that facilities could be made available to the merchant marine would be quite out of place. However, as a long-term idea this would be well worth looking at and could be of benefit to both sides.

ACKNOWLEDGEMENT

The author thanks the Ministry of Defence for granting him permission to present this paper. At the same time he wishes to point out that any opinions expressed are his own and do not necessarily correspond with Royal Navy policy.

APPENDIX 71.1

TYPICAL TRIALS PROGRAMME FOR A FRIGATE

General

Diesel generators: One day trial for each machine fitted.

Refrigerating and air-conditioning machinery: One day per machine plus one day for protection devices.

Machinery controls, 'A' series: Five days checking cold calibrations and system installation.

Part I

Machinery controls, 'B' series: Undertaken about four weeks after 'A' series and just prior to basin trial. Four days commissioning systems, 'hot' harbour conditions of load available.

First machinery installation inspection: Takes three days immediately prior to basin trial.

Basin trial: Three days' duration and includes harbour trials of distilling plant, steering gear, and stabilizers.

Second machinery installation inspection: Takes two days immediately prior to C.S.Ts.

Part II

Contractor's sea trials: Twelve days' duration (see under 'Special commissioning trials programme'); includes machinery controls, 'C' series, i.e. dynamic tuning under sea conditions and pen-recording datum plant conditions 'as new' for subsequent maintenance performance checking. Takes place four to six weeks after basin trial.

Opening up machinery for inspection: Items to be opened up decided by M.T.U. immediately after C.S.Ts. Includes boilers, which are inspected both on opening up and after external cleaning.

Part III

Third machinery installation inspection: Takes one day, immediately prior to final machinery trial.

Final machinery trial: One day duration, includes 2 h at full power and check of auxiliary machinery. Takes place four to six weeks before terminal date.

Terminal date inspection: Date by which all work, ex cleaning and painting, is to be completed. Takes half-day in harbour. Inspection undertaken by Commodore Superintendent, Contract Built Ships, with M.T.U. officer as technical adviser.

Final inspection: Date by which all work completed. Inspection is as for terminal date and takes place 14 days after.

Acceptance demonstration: Half-day duration, includes $\frac{1}{4}$ h at full power and check of steering gear.

Acceptance: Usually next day after acceptance demonstration.

TYPICAL PROGRAMME FOR CONTRACTOR'S SEA TRIALS

Day 1, at sea

Compass adjusting, anchor and capstan trials.

Check torsion meters (carried out daily on C.S.Ts).

Preliminary check of operation of steering gear, evaporators, stabilizers, feed regulators, and exhaust system controls.

Commence setting up forced lubrication systems.

Check auxiliary circulating water.

Demonstrate ability to change over boiler feed arrangements at 20 per cent power in both local and remote control.

Day 2, at sea
Demonstrate ability to change over boiler feed arrangements at 50 and 75 per cent powers in both local and remote control.
　　Stabilizer trials.
　　(Overnight, on return to harbour, check main gearing and freedom of flexible couplings.)

Day 3, at sea
Demonstrate ability to change over auxiliaries at 50 per cent power.
　　Maximum firing rate on each boiler in turn.

Day 4, at sea
Feed regulator and closed exhaust performance trials. High pressure–low pressure saturated reducing-valve performance trial.
　　Hand steering trial.

Day 5, in harbour
Fuel ship.
　　Check main gearing and freedom of movement of flexible couplings.
　　Sight boiler-furnace brickwork.
　　Make good defects.

Day 6, at sea
Twelve-hour consumption trial at 12 knots.
　　Evaporator trials, 12 h on closed exhaust.

Day 7, at sea
Stand-by trial (30 min stopped with steam on main engines), followed by running ahead at 180 rev/min.
　　Astern steering.
　　Full power astern for 15 min.
　　Functional trial ahead and astern (100 rev/min rapid manoeuvring each way for 30 min).

Day 8, at sea
Preliminary full power.
　　Bypass feed heater at full power.
　　Ahead steering.
　　Crash stop from full power.

Day 9, in harbour
Fuel ship.
　　Check main gearing and freedom of movement of flexible couplings.
　　Sight boiler brickwork.
　　Clean all boiler sprayers.
　　Make good defects.

Day 10, at sea
Six hours at full power.

Day 11
Spare day.

Day 12, at sea
Evaporator trials, 12 h on live steam.

COMMISSIONING OF LARGE POWER STATIONS

F. WALKER*

This paper gives a definition of commissioning philosophy, and presents a summary of commissioning procedures, organization, documentation, and personnel requirements. Some experiences and problems are noted concerning the commissioning of three South of Scotland Electricity Board power stations, conventional and nuclear. Special reference is made to chemical cleaning, steam purging, and condensate polishing followed by comments on packaged boilers, instrumentation and control equipment, load rejection tests, and special instrumentation. Finally, a brief mention is given on communications, operating intructions, and safety, and conclusions are drawn for present and future commissioning.

INTRODUCTION

THE INCREASING SIZE of boilers and turbo-generators, together with delays in bringing new plant on line, have focused considerable attention on commissioning techniques to achieve a safe and speedy transition from completion of construction to routine operation with reliable and satisfactory performance. Some commissioning techniques and experiences are examined, and possible improvements are noted with emphasis on boilers, turbines, and feed systems.

COMMISSIONING PHILOSOPHY

Power plant commissioning starts when any plant item operates for the first time, and is completed when the station is operating normally and routinely. There are three main stages:

(1) *Pre-commissioning.* Completion of construction, testing, and accepting plant items ranging from a motor/pump unit to a circulating water system or a switchboard to a complete control system.

(2) *Start-up.* Commissioning groups of items which form a unit, e.g. boiler or turbo-generator.

(3) *Full power operation.* Working complete plant units up to full power and routine operation; setting up and optimization of auto-control systems and boiler plant. This stage includes trouble-shooting and teething troubles and may extend to two years from start-up.

COMMISSIONING PROCEDURE AND ORGANIZATION

The S.S.E.B. Generation Operation Division are responsible, through the power station manager, for all plant

The MS. of this paper was received at the Institution on 6th October 1970 and accepted for publication on 16th November 1970. 32
* Developme and Commissioning Engineer, South of Scotland Electricity Board, Cathcart House, Inverlair Avenue, Glasgow S.4.

commissioning and the acceptance of plant for operation, the functions being controlled for the division chief engineer by the Board's commissioning engineer working in co-operation with the station manager and his staff and assisted by the station start-up engineer. This engineer is seconded to the station to be responsible for all commissioning and to lead a team of engineers posted to the station as temporary commissioning staff.

Organization for a multi-group contract

The commissioning organization varies between a multi-group and group, or consortium, contract. Procedures also vary between a conventional and nuclear station. Fig. 72.1 shows the commissioning organization for a conventional multi-group contract power station (Longannet). The organization involves S.S.E.B. projects and operations staff, consultants, and main contractors. The consultants are responsible for all contractors and normally speak for them at the commissioning committee. The committee is chaired by the station manager and directs the overall commissioning of the station. The requirements of the committee are carried out by a pre-commissioning panel and a start-up panel who are responsible for producing programmes and documentation and for the day-to-day commissioning work through test teams. Recently the two panels have been combined into one commissioning panel chaired by the start-up engineer. The committee meet monthly and the panel meet daily.

Organization for a group contract

Fig. 72.2 shows the commissioning organization for Hunterston 'B' nuclear power station. Three groups are involved: S.S.E.B. projects and operations staff and the group contractor who is responsible for all subcontractors on site. The commissioning committee is chaired by the

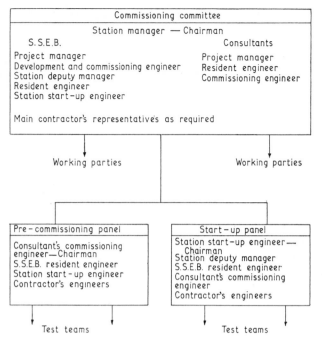

Fig. 72.1. Longannet power station commissioning organization

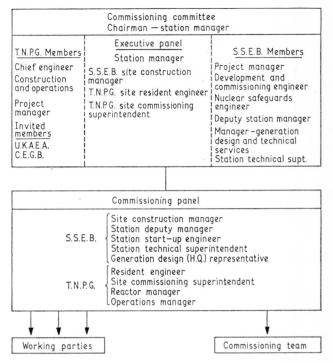

Fig. 72.2. Hunterston 'B' nuclear power station commissioning organization

station manager and is responsible for the overall direction of all phases of commissioning. Four site members form an executive panel (a quorum) readily available for major decisions and approval of programme changes, thus avoiding delays when commissioning is in progress.

The committee is supported by a commissioning panel chaired by the S.S.E.B. construction manager, prior to fuel loading the first reactor, and by the station deputy manager from fuel loading onwards. The panel, together with working parties, organizes the detailed work of commissioning within the principles and programmes approved by the committee. Actual commissioning is carried out on a day-to-day basis by the commissioning team (shown in Fig. 72.3), led by the group commissioning superintendent, and comprising contractors and seconded station personnel. Station staff operate all plant as agents for the contractor and to the requirements of the team. The station manager is responsible, through his staff, for safety during commissioning and has overriding control for nuclear safety.

At Hunterston 'B', as for all nuclear power stations, the Inspectorate of Nuclear Installations (through the Secretary of State for Scotland) must approve all arrangements for commissioning and must approve each stage on completion before the next stage can be commenced.

DOCUMENTATION AND PLANNING

The overall control of commissioning is specified by a commissioning manual, a comprehensive document produced for each station. For Hunterston 'B' this covers:

Commissioning procedure, staff structure, and responsibilities.

Commissioning programmes.

Testing procedures and documents.

Certification procedures for safety, operation, and maintenance.

Safety and permit-to-work systems.

Procedures for plant maintenance and defect clearance.

Routing of commissioning documents.

The manual was agreed and produced jointly by the S.S.E.B. and contractor's staff. It is given a wide circulation and the section on safety and permit-to-work systems is distributed to all subcontractors, site engineers, and others concerned with permit systems.

COMMISSIONING PERSONNEL

As power plant has become larger and more complex the permanent technical staff per megawatt have decreased in number, but more technical staff are required for commissioning. For a period of three years, 30 or more extra technical staff are required, both for operation and maintenance, to cover all stages of commissioning. Apart from actual engineering a large amount of paperwork has to be processed, not least being permits-to-work. Safety precautions require vigilance and continual monitoring by staff, especially with large numbers of contractors' personnel about the plant. Automatic and remote control facilities are seldom available in early stages, and operation

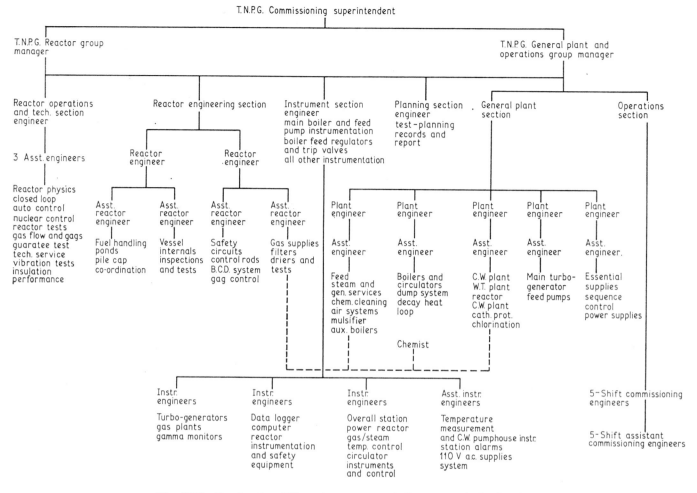

Fig. 72.3. Hunterston 'B' nuclear power station commissioning team

and checking equipment, coupled with the vast physical size of a modern station, absorbs extra staff.

The start-up engineer and commissioning staff should be on site well ahead of pre-commissioning, and ideally the senior engineers should have spent six months with design and project groups to become familiar with design philosophy and planning. If brought into the project sufficiently early, they could make a valuable contribution to the overall project. The same remarks apply equally to the permanent station staff.

At Longannet, it had been necessary to recruit a commissioning staff of 40, including the start-up engineer, and at Hunterston 'B' it is intended to appoint up to 30 commissioning staff, the start-up engineer being already in post. New stations in the S.S.E.B. commission at two-yearly intervals and build-up of commissioning staff can be achieved by transfer from a previous station, secondment from early appointments to a future station, and by direct appointments as commissioning staff. About one-third of the commissioning staff at a station will be finally absorbed by appointments to permanent staff or to other

operating stations, and it is valuable to have a reservoir of experienced engineers for this purpose.

COMMISSIONING EXPERIENCE AT S.S.E.B. POWER STATIONS

Past, present, and future commissioning are discussed for the three stations whose main parameters are shown in Table 72.1.

Cockenzie power station

The four 300-MW units were commissioned and fully operational between June 1967 and January 1969; times from synchronizing to full load varied from 10 months on No. 2 unit to 27 days on No. 4 unit, which subsequently achieved the best availability and highest load factor for similar units in the U.K.

Considerable completion delays were experienced: on No. 1 unit the boiler drum failure in 1966 and superheater failure during steam purging required replacements being taken from No. 4 boiler. During early operation the power

Table 72.1. Data for South of Scotland Electricity Board power stations

Particulars	Cockenzie	Longannet	Hunterston 'B'
Type	Conventional	Conventional	Nuclear A.G.R.
Fuel	Pulverized coal	Pulverized coal	Enriched uranium
Boiler plant			
Number of boilers	4 drum type	4 drum type	8 once-through 4 per reactor
Capacity, kip/h	2050	4000	950 each 3800 per reactor
Stop valve:			
Pressure, lbf/in²	2450	2450	2450
Temperature, °C	568	568	542
Reheater:			
Outlet pressure, lbf/in²	572	464	590
Outlet temperature, °C	568	568	541
Steam flow, kip/h	1610	2850	860 each 3440 per reactor
Final feed temperature °C	252	273	156
Turbo-generator			
Number of units	4	4 cross-compound	2
Continuous rating, MW	300	600	655
Speed, rev/min	3000	3000	3000
Stop valve:			
Pressure, lbf/in²	2350	2350	2350
Temperature, °C	566	566	538
Reheat:			
Pressure, lbf/in²	533	444	566
Temperature, °C	566	566	538
First unit synchronized	13th June 1967	22nd January 1970	December 1972
All units on load	17th January 1969	December 1971	August 1973

plant developed significant defects resulting in operational limitations, the worst of these being extensive cracking of the boiler drum stubs on units 1, 2, and 3. During stress relieving of No. 1 boiler drum, after welding modified stubs in early 1969, the drum end was overheated and the unit returned to service with restricted loading. A new drum was fitted *in situ* during summer 1970.

During commissioning, high silica levels on units 1, 2, and 3 necessitated long periods of low load running. The use of condensate polishing markedly improved No. 4 unit times.

Longannet power station

A Longannet boiler is shown in Fig. 72.4, and Fig. 72.5 shows No. 1 cross-compound turbo-generator, including h.p. heaters and steam feed pump. Pre-commissioning on No. 1 unit and general services progressed during 1969, after chemical cleaning and steam purging steam-to-set was achieved on 7th January 1970, and the generator synchronized 15 days later. Fig. 72.6 shows the start-up of No. 1 unit from synchronizing to full load, which was achieved in four months, the major part of this time being taken up with outages for fault rectification. The unit is now at full load, auto-control equipment is being commissioned, and the boiler control optimized to obtain efficient and reliable performance during winter 1970–71.

Floating No. 1 boiler safety valves   During the attempts to commission the safety valves at full pressure, in December 1969, four furnace tubes failed. It was suspected that these were caused by inadequate circulation and it was decided to make circulation measurements when the valves were next set. To monitor and control operations, special measurements were made of fluid density and velocity, heat flux and metal temperatures, and all other essential parameters including main feed valve regulating positions were recorded.

During March 1970 all safety valves were set at full pressure without water tube failures and with no physical change to the boiler. Density measurements in downcomer fluid enabled the adequacy of circulation to be assessed and proved a valuable 'driving' parameter, operational precautions being taken to avoid tube failures. Similar methods will be used to float safety valves on the other boilers, and are recommended for all large high-pressure boilers. It is considered that downcomer density

Key to Fig. 72.4

1. Drum.
4. First-stage radiant platen superheater.
5. Second-stage radiant platen superheater.
6. Secondary superheater.
7. Secondary reheater.
8. Primary reheater.
9. Economizer.
10. Primary superheater.
18. Forced-draught fan.
20. Main air heater.
45. Intervane burners.
49. Main steam outlet.
50. Primary reheater inlet.
51. Secondary reheater outlet.

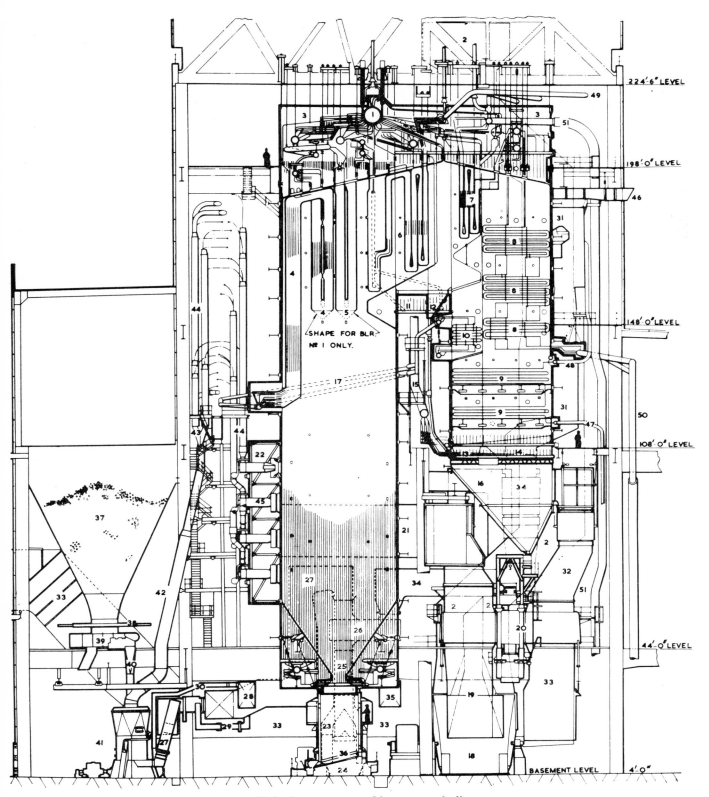

Fig. 72.4. Arrangement of Longannet boiler

Fig. 72.5. Longannet power station—No. 1 unit cross-compound turbo-generator h.p. feed heaters and steam turbo feed pump

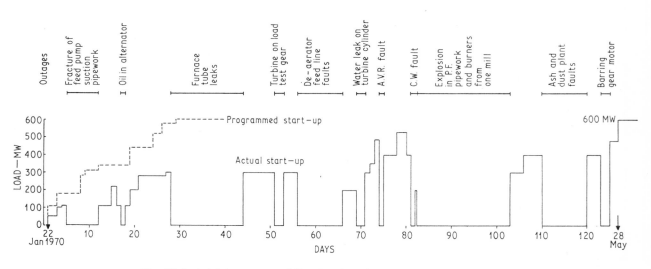

Fig. 72.6. Initial start-up of No. 1 unit at Longannet power station

measurements should be made permanent boiler instrumentation and there is a requirement for an instrument to give percentage steam-by-volume in a downcomer to the operator.

Special electrical testing Commissioning usually involves running the turbo-generator at low speed/low load for several days to carry out generator dry-out, protection tests, and automatic voltage regulator testing. Methods were developed and used on No. 1 unit to carry out these tests prior to steam-to-set, saving 15 days in line time and reducing low speed/load running to 12 h.

Generator dry-out The cross-compound generators were dried out by d.c. circulating current, a test dry-out being first carried out on one generator by circulating 4000 A in the main windings. During the test the machine was ventilated with dry air and had temporary thermocouples fitted in the end windings. After 120 h an acceptable insulation resistance was obtained and a relationship established between temporary and permanent thermocouples which eliminated the need for temporary thermocouples for programmed dry-outs.

The programmed dry-out of both generators was done using motor generators for the h.p. generator and a static rectifier for the i.p. generator. Circulating currents were limited to 2500 A and the dry-out was completed in 250 h.

Generator unit protection testing This was carried out by supplying 11 kV to the 275 kV side of the generator transformer with temporary connections across the generator windings. Test currents of 5500 A were obtained in the generator circuits and out-of-balance currents were extrapolated for full load values. Synchronizing equipment was checked by back-energizing the generator and unit transformers at 275 kV.

Automatic voltage regulator testing The output from a motor-driven high-frequency generator, from No. 3 unit, was fed to the output terminals of No. 1 unit h.f. generators to enable the A.V.Rs to be tested and commissioned. The test revealed several faults which could have caused delays on a running unit.

General interest Partly due to oil heater faults, the boiler commissioning was carried out using 35-s oil instead of normal light-up 950-s oil. This generally assisted commissioning.

P.f. mill fires are a problem and a system is being developed to sample CO from mills and record changes to give advanced fire warning.

Early and thorough pre-commissioning of all plant items eased problems during start-up. Even so, more could have been done, especially on coal and ash plants. It is important to pre-commission all items and services

Table 72.2

Test details	Test stages	Days
Combined engineering tests. Reactor vessel circuits, blowers, and associated plant	01–54	170
Reactor physics tests	55–59	16
Preparations for and chemical cleaning, using blower heating	60–62	10
Initial raising power to 20 per cent, using dump condenser	63–75	25
Initial raising power from 20 to 100 per cent. Steam to turbine at Stage 82, the turbine will have been pre-commissioned on steam from the 'A' station	76–136	53
Tests at 100 per cent power and subsequent tests. Includes acceptance tests, load reject tests, and reactor trip tests	137–183	30
Total time to reach 100 per cent load	—	274
Total time to complete all tests	—	304

early, carefully, and under conditions met during operation.

Delays and shutdowns were caused by faulty cable terminations and interference with terminations during commissioning, and emphasizes the need for better design and segregation of main terminal racks with adequate security during operations.

Hunterston 'B' nuclear power station

Two 660-MW A.G.R. units with once-through boilers are under construction and programmed for full commercial operation by February and August 1973, respectively. Commissioning of the first unit commences about February 1972 with pre-commissioning of plant items and services starting early 1971. A summary of programme times is shown in Table 72.2.

CHEMICAL CLEANING AND STEAM PURGING
Need for cleaning

Modern boilers and associated feed systems require a high standard of cleanliness, complete freedom from foreign matter, and a sound, uniform oxide film on boiler internal surfaces. This is essential to prevent on-load corrosion, turbine fouling, and damage to plant. Despite care taken during manufacture, pre-cleaning, and improvements in site storage and erection, plant must still be chemically cleaned and steam purged although these exercises are time-consuming and expensive. The standard procedures specified by the S.S.E.B. are:

Preliminary water flush with cold demineralized water.

Alkali degrease of main boiler, economizer, feed, and bled steam systems.

Acid clean (hydrochloric) of boiler and economizer, excluding superheater and reheater.

Acid clean (citric) of all systems.

Passivation of boiler, economizer, superheater, and reheater.

The chemical cleaning is done immediately prior to the boiler going on load and is concluded by steam purging all pipework between boiler and turbine. A steam blow should achieve a shifting factor between 1·5 and 2·0, this factor being the ratio of steam momentum during blow to momentum under full-load conditions. Steam purging is complete when reasonably clean etching plates are obtained.

Experience at Cockenzie power station

The standard procedure was followed successfully on all units. During No. 1 unit feed system clean, a degreasing solution was carried to the attemporator spray lines and not completely removed. Later, during steam purging, a leak in the superheater revealed numerous cracks in the austenitic tubes; this was eventually identified as stress corrosion cracking caused by a degreasing solution being injected into the attemporator and carried to the super-

heater. This incident emphasizes the need for strict control, inspection, and monitoring throughout all stages of chemical cleaning.

Experience at Longannet power station

No. 1 unit has been commissioned and cleaning is completed on No. 2 unit. Originally it was hoped that high standards during manufacture, storage, and erection would have reduced feed system cleaning to an alkaline degrease, but in the event comprehensive cleaning of all systems has been required. Fig. 72.4 shows the boiler arrangement.

On No. 1 unit incomplete boiler lagging prevented alkali boil-out, but an alkali wash circulated at 95°C gave good results. The boiler and economizer were acid cleaned with hydrochloric and citric—the superheater and reheater, having austenitic sections, being only citric cleaned. Heating was by the installed boiler burners.

On No. 2 unit oil burners were not available and cleaning was carried out using packaged boiler heating,

Fig. 72.7. Longannet power station—No. 1 unit h.p. line turbine with temporary steam purge pipework fitted to valve chests

electrical resistance heating being considered as an alternative. The superheater and reheater austenitic sections were not chemically cleaned, as difficulty had been experienced on No. 1 boiler in flushing these sections, and it was considered that steam purging would give adequate cleaning. The overall time for cleaning No. 2 unit was 19 days (only 4 in line) compared with 29 days in line time for No. 1 unit.

Steam purging on No. 1 unit produced high shifting factors, up to 1·95, partly due to starting blows with very low drum water levels. Only 26 blows were needed and target etching plates were fitted when blows were calculated to have been effective. Two plates were fitted for each system (superheater and reheater) and were satisfactory. Fig. 72.7 shows the h.p. line turbine with temporary steam purge pipework fitted to the valve chests. On units 3 and 4 modifications will allow these connections to fit on to the steam strainers which will save six weeks' reinstatement time and allow completion of turbines and pre-start checks to be done prior to start-up.

Hunterston 'B' cleaning procedures

It is intended to restrict cleaning of the once-through boilers to a citric acid clean and passivation, and to clean all other circuits by alkali flushing and a citric acid clean. Chemical cleaning will be carried out prior to reactor start-up using gas circulator heating.

All pipework will be maintained at a high standard of cleanliness throughout manufacture, storage, and erection to reduce the amount of cleaning required but it is intended to carry out steam purging prior to chemical cleaning using steam from the 'A' station.

Summarizing the cleaning of large units

The design must cater for temporary connections for cleaning.

Chemical cleaning should not be necessary for superheater and reheater sections.

A circulated alkaline wash is equivalent to an alkaline boil-out.

Cleaning can be done out of line using alternative heating.

Demineralized water must be used for all operations.

Strict precautions must be taken to keep stress-corrosion-raising chemicals away from austenitic sections.

Adequate pumping capacity must be available to ensure high circulation rates and thorough flushing.

CONDENSATE POLISHING PLANT

With high-pressure boiler plant, particularly once-through boilers, the feed water quality is of prime importance. The most critical period for plant is during commissioning and early operation when poor start-up and inadequate chemical control will result in contamination of the boiler internal surfaces. Unless condensate polishing plant is used units are restricted in load for long periods during start-up, and vast quantities of water are dumped until

Table 72.3. Water quality from condensate polishing plant

Type of boiler	Contaminant (maximum p.p.m.)				Conductivity, μmho/cm
	Iron	Copper	Total silica	Sodium	
Drum type .	0·01	0·005	0·015	0·01	0·10
Once-through .	0·005	0·0015	0·015	0·005	0·08

satisfactory conditions are achieved. For large high-pressure plants the quality of water required from a polishing plant is shown in Table 72.3.

During turbine start-up the time taken to recover water to the boiler is reduced to one-third by using condensate polishing and, in addition, a systematic clean-up of the feed system can be carried out prior to start-up by recirculating water through the polishing plant. As most modern plant will eventually have to be two-shifted, a condensate polishing plant is essential and has the advantage that large-capacity polishing plant can be continually in circuit to deal with small condenser leaks.

In the U.K. the best plant arrangement for condensate polishing has not yet emerged. The possibilities are at least six variations of filter–cation–mixed bed units and a filter system using powdered resin.

Experience with condensate polishing

At Cockenzie the polishing plant was not installed and the poor quality of feed water when commissioning units 1, 2, and 3 resulted in protracted low load running with condensate discharged to waste until the system was cleaned up. For commissioning No. 4 unit an experimental condensate polishing plant (filter using powdered resin) was installed. Although only handling 10 per cent full flow, the plant enabled condensate to be recovered to the boiler in 44 h on initial start-up compared to 143 h on No. 2 unit.

For Longannet, condensate polishing was installed as part of the normal feed water make-up plant by increasing the capacity of the mixed bed units to allow 20 per cent polishing for any one boiler/turbine unit. The arrangement also allows pre-start clean-up of the feed system. The effectiveness of the plant is shown by the graph in Fig. 72.8 of iron levels for the start-up of No. 1 unit. Resulting from the Cockenzie experiment a portable 50 per cent duty condensate polishing plant (filter using powdered resin) has been obtained for commissioning units 2, 3, and 4.

At Hunterston 'B', with once-through boilers, 100 per cent condensate polishing is essential. The plant will be separate from the feed water make-up plant and will comprise pre-coat candle filters, cation, and mixed bed units.

PACKAGED BOILERS

Oil fired packaged boilers are standard commissioning equipment, and the trend is now to install them as permanent auxiliaries in addition to commissioning duties.

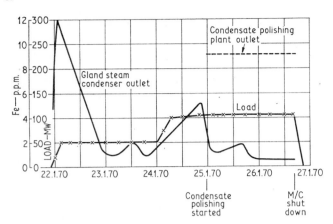

Fig. 72.8. Condensate polishing plant outlet during initial start-up of No. 1 unit at Longannet power station

During commissioning, packaged boilers may be used to supply heat for chemical cleaning and steam purging; for deaerator heating to maintain low oxygen levels; and for turbine pre-commissioning up to speed and synchronizing.

For nuclear systems, e.g. S.G.H.W.R., using active steam, packaged boilers may be used to commission turbine plant to avoid radioactive working during checking and fault rectification.

It is claimed that packaged boilers have saved up to eight weeks' time during commissioning at certain stations.

S.S.E.B. experience

At Cockenzie two package boilers were installed for commissioning, chemical cleaning, and deaeration. They have been retained as permanent auxiliaries. Three package boilers are in use at Longannet for commissioning use only for chemical cleaning of boilers and feed systems. Hunterston 'B' will have two package boilers for permanent auxiliaries, and commissioning steam will be supplied by the 'A' station.

CONTROL SYSTEMS AND INSTRUMENTATION

Modern boiler/turbo-generator units, auxiliaries, and systems have grown rapidly in size, technological advance, and complexity. This has been accompanied by a similar expansion in the control systems and instrumentation. The large physical size of power stations, coupled with reduced numbers of operating staff, has resulted in the majority of equipment being controlled from a central control room with complex auto-control and interlocking schemes installed throughout. (At the same time alternative 'local to plant' control and indication has decreased considerably. Hence, the successful and early commissioning of remote control and indication systems and instrumentation is essential for safe and efficient operation of plant.) Recent experiences in the U.K. where plant has been damaged and availability lost due to operating without full control, indication and alarm facilities have

emphasized the necessity for all these facilities to be commissioned before plant is operated. Both the Electricity Boards and manufacturers decree that no plant should be run without these facilities, but to date no unit over 200 MW has been synchronized initially with full control and indication facilities available. Control and instrumentation systems are always the last items available for completion and commissioning, and invariably with the plant well behind programme and the system hard pressed for load the main plant items are put into operation without allowing time to complete control and instrumentation commissioning.

Neither Cockenzie nor Longannet power stations were exceptions to this state of affairs. All units at Cockenzie were commissioned with instrumentation and control systems not fully available or unreliable, and though some improvement was made by the time No. 4 unit was commissioned the general position was unsatisfactory and would have been worse if the S.S.E.B. had not put in considerable effort with its instrument and engineering staffs to assist overworked contractors.

At Longannet power station the Cockenzie experience had been noted and improvements were expected. In the event twelve S.S.E.B. instrument engineers had to be used for several months to assist contractors with control and instrumentation systems to enable No. 1 unit to be synchronized with enough facilities available for safe operation. To date the turbine automatic control equipment and the auxiliary preparation equipment has still to be commissioned. The boiler auto-control equipment is only partly commissioned and the data logger equipment not fully operational.

Results achieved at Longannet No. 1 unit was first synchronized on 22nd January 1970 with the following remote control and instrumentation available:

	Items commissioned
Remote control:	346 out of 433
Instruments and indication:	650 out of 830
Alarms:	283 out of 522

One hundred per cent should have been achieved by this time.

The results are better than at Cockenzie power station or other large units previously commissioned in the U.K. but are still not satisfactory. In the commissioning stages prior to synchronizing No. 1 unit at Longannet, and particularly the chemical clean and steam purge stages, the instrumentation and control position was most unsatisfactory and caused programme delays. The experience on No. 1 unit was fully analysed and the lessons learnt have been applied to No. 2 unit where a great improvement has been obtained, and it is expected to synchronize with 100 per cent availability of instrument and control systems, excluding auto-control systems which can only be commissioned as the unit works up to full load. The use of an instrumentation and control working party, led by an S.S.E.B. engineer, has had a marked effect on progress.

Summary Failure to obtain 100 per cent availability of control and instrument equipment is mainly due to (1) late erection of plant and late completion of cabling; (2) delay in finalizing designs and the large number of modifications involved; and (3) underestimation of the work content in engineering and commissioning the systems.

Apart from fulfilling the above requirements, the following are necessary to ensure that equipment is proved before it is operationally commissioned:

All equipment should be fully tested at works.

Pre-commissioning on site should be as extensive as possible.

Special provision must be made for early completion of air and electrical supplies.

Boiler simulation studies are necessary to prove the logic and viability of control systems and to establish controller settings.

Nuclear power station control and instrumentation

Generally the instrumentation and control position at nuclear power stations has to date been considerably better than at conventional power stations, the reasons probably being two-fold: a firm belief by all concerned that instrumentation and control systems really are essential before commissioning, and the stimulation of the Inspectorate of Nuclear Installations' requirements for safe operation. Whether or not this situation will exist at the new A.G.R. power stations now under construction, with their more complicated control and computer systems, remains to be seen but the above recommendations are equally applicable.

LOAD REJECTION TESTS

As part of the commissioning tests at least one turbo-generator unit in a new station should undergo a full load rejection test for the following reasons:

(1) To prove correct operation of the turbine governor system and steam admission valves.

(2) To observe the behaviour of the turbine, boiler, and auxiliary plant.

(3) To check if the turbo-generator, after a load rejection, can remain at normal speed supplying its own auxiliaries from the unit transformer and for what period it can do this. Further, to check the feasibility of resynchronizing the generator and rapidly reloading the unit.

(4) To give the operating staff experience and confidence in handling the situation.

Prior to carrying out a load rejection test, very thorough checking of all systems is essential. Extra recording instrumentation is required and operating personnel briefed. The turbo-generator must be tripped and over-speed tested initially deloaded and running light and load rejections carried out at part load before attempting a full load test.

Tests at S.S.E.B. new stations

Load rejection tests have been carried out at Cockenzie up to two-thirds full load. The machine tended to exceed the designed governor transient of 9·5 per cent due to reheat interceptor valves passing and the governor not meeting its design characteristic. Further tests will be done when these are satisfactory and as a temporary measure the overspeed trips are set to 7·5 per cent. It is intended to carry out load rejection tests on Longannet No. 1 unit before the end of 1971. The commissioning programme at Hunterston 'B' includes load rejection tests on one machine at 25, 75, and 100 per cent full load.

A test was carried out at Kincardine power station recently when a 200-MW unit had a simulated unplanned 50 per cent load rejection to assess the machine performance when supplying its own auxiliaries and without staff being briefed. Although carried out under adverse conditions the test was successful, the unit proving that it could supply its own auxiliaries for at least 90 min after a load rejection.

SPECIAL INSTRUMENTATION DURING COMMISSIONING

Turbines

For commissioning No. 1 turbine at Longannet power station the contractors fitted additional instrumentation to the turbine to give advanced warning of adverse trends in the following:

High-pressure cylinder—A 10-slave manometric system was fitted to monitor internal radial clearance changes caused by outer casing vertical deflection.

Intermediate pressure No. 1 cylinder casing—A six-slave manometric system was fitted to the i.p. No. 1 casing lower half flange to monitor internal clearance.

Gland clearances—These were measured using air gauges mounted on the outside of the gland covers. On the h.p., i.p. No. 1, and i.p. No. 2 cylinders the gland clearances were measured at the top, bottom, and both sides. On the l.p. cylinders only top clearances were measured on each line.

Summarizing the results obtained and the usefulness of this special instrumentation:

High-pressure cylinder—The internal radial clearances were satisfactory during steady running conditions, but after shutdown a high hog develops resulting in minimum internal clearances on certain diaphragms. Two incidents of high eccentricity have indicated the sensitivity of the cylinder to changes in internal clearance. Modifications are being carried out which include lagging and sole plate arrangements, but while these will effect an improvement, continued monitoring of critical parameters must continue.

Intermediate pressure No. 1 cylinder casing—A slight casing hog was measured during normal on-load running and on shutdown when on barring gear. With the maximum casing distortion measured, the minimum radial clearances were adequate for cylinder safety.

Gland clearances—During steady running the gland clearances have been satisfactory. Following shutdowns early in commissioning, certain side clearances were seriously reduced but modifications to the transverse cylinder location have removed this problem. Other top clearances may become critical during a cold start if the rate of reheat temperature rise is not carefully controlled. One incident of abnormal, but not critical, top clearance has been recorded. Investigations into clearance changes continue and similar measurements will be carried out on No. 2 unit.

On No. 1 turbine at Longannet the value of special instrumentation during commissioning was proved as indications were given in time for corrective action to be taken to avert what might have been a serious failure. The use of special instrumentation will continue on No. 2 and subsequent units, and most of the temporary instrumentation is being retained on No. 1 unit. Limiting parameters and strict operating procedures have been established for No. 1 unit and consideration is being given to fitting additional permanent instrumentation to both No. 1 turbine and certain auxiliary plant. It has also been recognized that hazards can exist when a machine is shut down arising from defective plant external to the turbine. Finally, much information has been obtained which is being carefully studied by the turbine designers.

Boilers

Special instrumentation has been fitted to one boiler at both Cockenzie and Longannet power stations to ensure efficient commissioning and to establish safe operating procedures leading to high availability of all boilers.

Instruments were fitted to give information on furnace wall heat absorption rates and temperatures, superheater and reheater tube and steam temperatures, circulating water densities and velocities, flue gas temperature and analysis, and feed water chemistry. Heat absorption rates were measured by heat flux meters, tube metal temperatures by stainless steel sheathed thermocouples, and steam temperatures measured by thermocouples on the outside of tubes clear of gas passes. Circulating water velocities were measured by pitot tubes and water density determined by differential pressure head measurement. At Cockenzie 590 special instruments were fitted and together with other normal instrumentation the outputs were fed to a data logger, the results being produced on punched tape which could be printed out or analysed by computer. Certain outputs were displayed visually. On commissioning the boiler, comprehensive tests were carried out to observe the effects of load, firing pattern, excess air, feed water temperature, blowdown, drum water level, etc.

Steady-state tests took about 2 h, computer analysis was obtained within 24 h, and a test report issued three days later. There were many instances where the special instrumentation proved invaluable in detecting problems,

amongst these was the early detection of high tube metal temperatures at certain operating conditions on superheater radiant platen tubes and the establishment of operating techniques to prevent overheating.

COMMUNICATIONS

During commissioning of power station plant the communication system between personnel directly engaged has, to date, not been satisfactory and has led to delays and frustration. Various forms of communications tried include (*a*) public address systems, where people for whom a message is intended do not hear it or get a confused version in noisy locations, (*b*) the station telephone system, which is inconvenient and not practicable in a noisy location or where persons are moving about, and (*c*) two-way radio-telephone which is limited to a few persons.

It is essential that every person in a commissioning team should be able to receive instructions and be kept up to date at all times with progress reports—especially if delays have occurred which require changes of programme. It is also vital to be able to transmit accurate instructions rapidly during an emergency. Briefly, all require to receive and a few to transmit and receive messages. A system is being investigated using u.h.f. radio which will allow a central station, normally in the central control room, to transmit to 30 persons and transmit and receive to six key personnel. Provision would be made for hearing-aid type speakers incorporated into ear defenders for noisy locations, and various types of microphones.

OPERATING AND MAINTENANCE INSTRUCTIONS

Operating instructions should be available prior to commissioning, especially for start-up, and maintenance instructions ready when plant is taken over. All detailed test schedules must refer to the appropriate operating instructions and if the finalized instructions are not available, which is often the case, then temporary instructions should be produced for commissioning.

Modern plant requires comprehensive instructions, which may cost over £250,000. The station may produce all instructions using information from contractors and technical writers or, in the case of a group contract, the contractor may produce the instructions, which are vetted by the station. The first method involves the station technical staff in 50–60 man years' work and the second in 15–20 man years' work.

SAFETY DURING COMMISSIONING

The transitional period from 'plant under construction' to 'plant operational' can be an accident area as plant becomes alive under station control, safety rules, and permits-to-work and later is operated by station staff acting as contractors' agents until take-over. Large numbers of contractors' personnel are working on plant during this period, many lacking experience of permit systems, unconscious of possible dangers, and often with poor communications and control. The commissioning committee should institute, vet, and monitor all permit systems. They should ensure that contractors' personnel receive information and training on safety aspects and should insist that all commissioning documents include a safety section.

CONCLUSIONS

Commissioning techniques have progressed with plant development but improvements can be made in all areas. The major requirements are:

(1) An efficient commissioning committee and commissioning organization with early and thorough preparation of planning and documentation.

(2) The early appointment of adequate commissioning staff led by a competent start-up engineer.

(3) Plant initially designed to facilitate commissioning requirements.

(4) Detailed plant pre-commissioning under operating conditions and the development of ideas for commissioning plant in out-of-line time.

(5) Special instrumentation to assist commissioning and improve availability.

(6) Improvements in manufacture, storage, and erection of plant to simplify or remove the need to chemically clean and steam purge.

(7) The use of condensate polishing plant and packaged boilers to reduce time.

(8) The early completion of instrumentation and control systems, their full pre-commissioning proving by model or simulation and provision of ample staff to carry the work load.

(9) The establishment of a good team spirit between all concerned during commissioning.

ACKNOWLEDGEMENTS

Acknowledgements are made to the South of Scotland Electricity Board for permission to publish this paper and to the English Electric Company and the Nuclear Power Group for supplying information. The author's thanks are due to colleagues in the Board for their assistance.

REVIEW OF ELECTRIC POWER SUPPLY IN THE NETHERLANDS

G. A. L. VAN HOEK*

This paper gives a review of electricity supply from the standpoint of electricity generation and of the interconnection of power stations in the Netherlands. Some typical data of the electricity supply industry are given, and emphasis is laid on the fact that generation in the Netherlands is entirely on a thermal basis. Special requirements originating from this condition are mentioned and a co-ordinated and obligatory extension programme for generating plant is described. The paper stresses that production units form an integrating part of the electricity supply structure and thus must fulfil a task which, to a large extent, is defined by the variation of consumers' demand. Periodic site testing to verify that generating plant matches the system characteristics is strongly recommended.

INTRODUCTION

ELECTRICITY SUPPLY in the Netherlands is now a matter of provincial and municipal care, although, as in many other countries, it was started as a private enterprise.

As long ago as 1895 the first municipality—that of Rotterdam—founded an electricity undertaking, and this was followed between 1895 and 1910 by almost every important city and town. As there was then no fixed pattern for the development of these municipal undertakings they extended to the rural environment, and in some cases to neighbouring towns. A number of private companies also developed in some of the provinces.

This situation caused the provincial authorities some concern, and as they decided that the supply of electricity should mainly be a provincial task, provincial undertakings were founded. Some of these provincial undertakings had the status of public services; others, however, were founded as limited companies in which the shares were owned by the provinces.

Nowadays the distribution of electrical energy is in the hands of municipal or provincial authorities, and the production of electrical energy and rural electrification have generally become provincial tasks.

In the province of South Holland the six municipalities that originally owned power stations founded a limited company for the production of electricity, the shares of this company being in the hands of the province and of the six municipalities concerned. Four of these municipalities still have their own power stations, and the city of Amsterdam also has its own power station.

The provincial undertakings gradually absorbed private regional companies, although a few of these managed to survive for some time. In 1950 the last private electricity supply undertaking in the Netherlands was taken over by one of the provincial companies. Thus the total public electricity production in the Netherlands today is concentrated in 11 power stations or combinations of power stations:

(1) the Provinciaal Electriciteitsbedrijf in Friesland, at Leeuwarden;

(2) the Elektriciteitsbedrijf voor Groningen en Drenthe, at Groningen;

(3) the N.V. Electriciteits-Maatschappij Ijsselcentrale, at Harculo, Hengelo, and Zwolle;

(4) the N.V. Provinciale Geldersche Electriciteits-Maatschappij, at Nijmegen and Lelystad;

(5) the N.V. Provinciale Limburgse Electriciteits-Maatschappij, at Buggenum;

(6) the N.V. Provinciale Noordbrabantsche Electriciteits-Maatschappij, at Geertruidenberg;

(7) the power stations co-ordinated by the N.V. Electriciteitsbedrijf Zuid-Holland, at Leiden, Den Haag, Rotterdam, and Dordrecht;

(8) the Provinciaal Electriciteitsbedrijf van Noord-Holland, at Velsen;

(9) the city of Amsterdam, in Amsterdam;

(10) the N.V. Provinciaal en Gemeentelijk Utrechts Stroomleveringsbedrijf, at Utrecht;

(11) the N.V. Provinciale Zeeuwse Energie-Maatschappij, at Middelburg and Vlissingen.

In addition, there are 86 municipal undertakings which attend to the distribution of electrical energy.

Shortly after the second world war the decision was

The MS. of this paper was received at the Institution on 7th December 1970 and accepted for publication on 31st December 1970. 33
* Manager, N.V. Samenwerkende Electriciteits-Productiebedrijven, Utrechtseweg 310, Arnhem, Netherlands.

made to extend the high-voltage lines which existed between some regional power stations, to make a grid that would interconnect almost all the power stations or groups of power stations mentioned above. The 10 large electricity supply companies concerned at that time established a limited company, the N.V. Samenwerkende Electriciteits-Productiebedrijven (SEP), in order to promote the co-operation of the Dutch electricity supply companies via the grid. The Provinciale Zeeuwse Energie-Maatschappij, whose power stations were less important at that time, was not yet to be directly connected with this grid but it was tied to the nearest big power station by means of a 50-kV line. Thus, the installed capacity of the power stations in Zeeland also contributed to the interconnected power.

It was decided that the grid voltage would be 150 kV in the central, southern, and western parts of the country and 110 kV in the north-east, while interconnection with Germany and Belgium took place at 220 kV.

The lines and the necessary transformer stations in the grid remained the property of the electricity generating companies, and the lines interconnecting regional companies became the common property of the two electricity companies concerned, each bearing 50 per cent of the expenses.

All technical decisions to form a grid out of the existing high-voltage lines were taken by a technical committee consisting of representatives of the 10 partners and SEP. This committee dealt with such subjects as insulation level, earthing of the neutral, technical specifications, layout of the telecommunication network, and so on.

Each of the 10 companies established a regional dispatch centre, and SEP set up a national system control centre at Arnhem. These centres work in close collaboration to ensure the continuity of the electricity supply, i.e. the generation and transmission of electricity via the grid. They take precautions to operate at a sufficiently safe level of electricity supply, and they take action in emergencies to restore the supply as soon as possible. In recent years the first phase of a superimposed grid for 380 kV has been constructed, and the remote control of this 380-kV network has been added to the tasks of the national system control centre.

As a logical consequence of historic development this

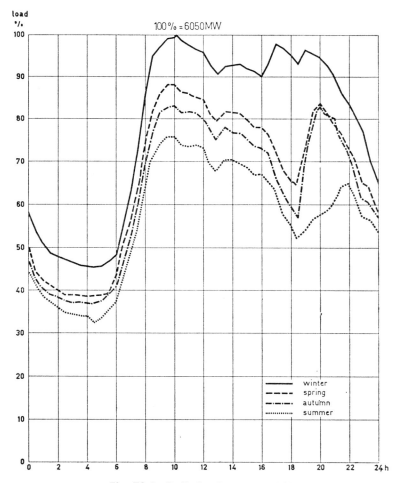

Fig. 73.1. Daily load curves, 1969

new grid has been built by SEP, with the collaboration of the 10 participating electricity generating companies. Together with the introduction of the new grid, a new interconnection with Germany (on the 380-kV level) was built and the decision to construct a new interconnection on the same level with Belgium was made soon afterwards.

HISTORIC AND RECENT DATA

The demand for electrical energy in the Netherlands has grown rapidly over the last 20 years. While peak demand in 1949 was only 1100 MW, this figure rose to 2525 MW in 1959 and attained 6050 MW in 1969. The rise in the last decade was equivalent to a doubling of the peak load in a period of eight years. The last few years show even higher rates of growth: 11·8 per cent in 1968 and 10·4 per cent in 1969, compared with the highest peak load in the preceding year.

The energy requirements in megawatt-hours showed an even more marked increase: this item rose from 4 600 000 MWh in 1949 to 11 900 000 MWh in 1959, and attained 31 550 000 MWh in 1969—which indicates that consumption doubled in a period of seven years—while a rise of 13·6 and 12·2 per cent was evidenced in 1968 and 1969 respectively, as compared with the preceding year.

For many years the Netherlands has been a modestly industrialized country. That this has been changing rapidly of late is reflected in the relative increases in power demand during recent years. For the next eight years an increase of 9·5 per cent per year for the highest peak load and of 10·5 per cent per year for the energy requirement are estimated as average values.

In Fig. 73.1 daily load curves for normal working days are given for four typical seasons of the year. In Fig. 73.2 the load–duration curve for the year 1969 is shown as well as the equivalent peak load utilization period which is equal to the total demand in megawatt-hours divided by the peak load in megawatts. This peak load utilization period amounted to 5210 h in 1969. This item was subject to a marked increase of about 10 per cent during the last seven years, which is the period in which the total demand doubled. The increase in the load utilization factor of the consumers is also shown to some extent by the rise in the ratio between the monthly minimum night load and the highest peak load occurring in the same month. This ratio, as well as its progressive yearly average, has been given in Fig. 73.3 for the corresponding period.

The production of electricity in the Netherlands is characterized by the fact that no economic water power is available. The entire production is based on oil, natural gas, coal and, to a very small extent, nuclear energy. In Fig. 73.4 the relative consumption of these types of fuel has been registered for the last seven years, the total consumption taken as 100 per cent every year. The advance of the use of natural gas is apparent, as is the reduction of the consumption of coal. This trend will continue in the near future, and it is estimated that within the next seven years the consumption of natural gas in power stations will

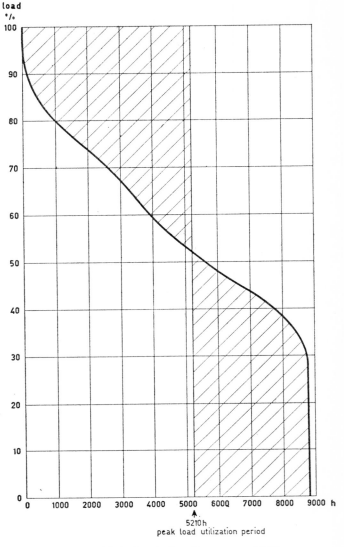

Fig. 73.2. Load–duration curve, 1969

reach 67 per cent of the total consumption of fuel for electricity production.

The total amount of generating plant installed at the end of 1969 was 8510 MW. Of this, 8 per cent consisted of new production equipment, a further 58 per cent varied in age between 1 and 10 years, 14 per cent was 11–15 years old, and the remainder—i.e. 20 per cent—was older than 15 years with much of this older than 20 years.

Until now only an insignificant part of the generating plant has consisted of gas turbines (about 2 per cent), but this will change as the gain in efficiency obtained by installing new large conventional units becomes less and less when unit sizes increase, and consequently the need for destining the original generating plant to peaking purposes will gradually decrease.

A study to determine the optimal composition of the generating plant mix in the near future is being

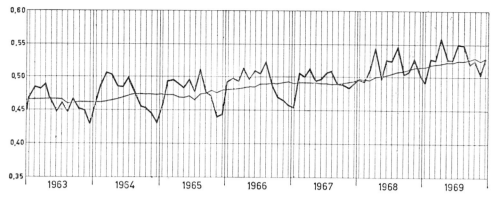

Fig. 73.3. Ratio between minimum night load and maximum peak load per month and progressive yearly average

undertaken in the Netherlands. The use of nuclear energy for the production of electricity has not yet been regarded as competitive for the unit sizes used in the Netherlands system.

Of course, the load utilization factors are taken into account. Only a very small nuclear power station, with a capacity of 54 MW, has been constructed in an attempt to gain experience in the operation and instrumentation of power stations of this kind. Further, a nuclear power station of 470 MW is at present under construction in a part

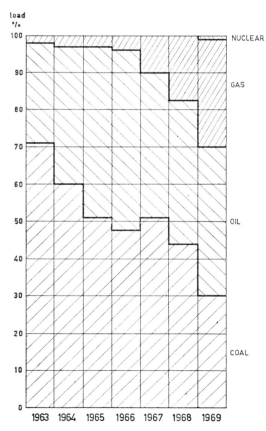

Fig. 73.4. Relative fuel consumption, 1963–69

of the country where a sudden rise in load, as well as in load utilization factor, has been caused by the introduction of new industries.

The largest units in service in the Netherlands now have a capacity of 230 MW, and 50 per cent of the installed capacity consists of units bigger than 100 MW. Until recently the unit size was restricted to a value of about 200 MW, as the transmission capacity available in the grid was not sufficient to back up sudden outages of generators of a greater capacity. The commissioning of the supergrid for 380 kV in 1970 lifted this restriction and the first generator with a capacity of 400 MW will enter service in 1971; it is expected that the first 500-MW set will be commissioned in 1974.

Today, the highest voltage in the central, southern, and western parts of the country is 380 kV, whereas in the north-east the highest voltage is 220 kV. These two networks have been superimposed on the original networks for 150 and 110 kV, which have been split up appropriately. The interconnection between the 380-kV network and the 220-kV network in the north-east of the country will be directly established in the course of 1972, but for the time being it is formed with a part of the 150 kV network as an intermediary.

Fig. 73.5 shows the outline of the Dutch grid and indicates the interconnections with the neighbouring countries. The shaded arrows in the 380-kV network indicate the intention to close the loop of 380-kV lines at some time in the future. Although not yet decided, this may take place between 1975 and 1980.

With a few minor exceptions, the whole grid consists of double circuit lines. These lines are fully equipped at both ends with switching apparatus and protective equipment, and special care has been taken to shield the lines against lightning to achieve a high standard of reliability.

Operation of the network is based on the single contingency principle, which also holds for the power generation. This means that the amount of spinning reserve and the loading of the network are such that neither a sudden outage of a big generator nor a sudden tripping of a circuit of the high-voltage network can cause an interruption of the supply to the consumers.

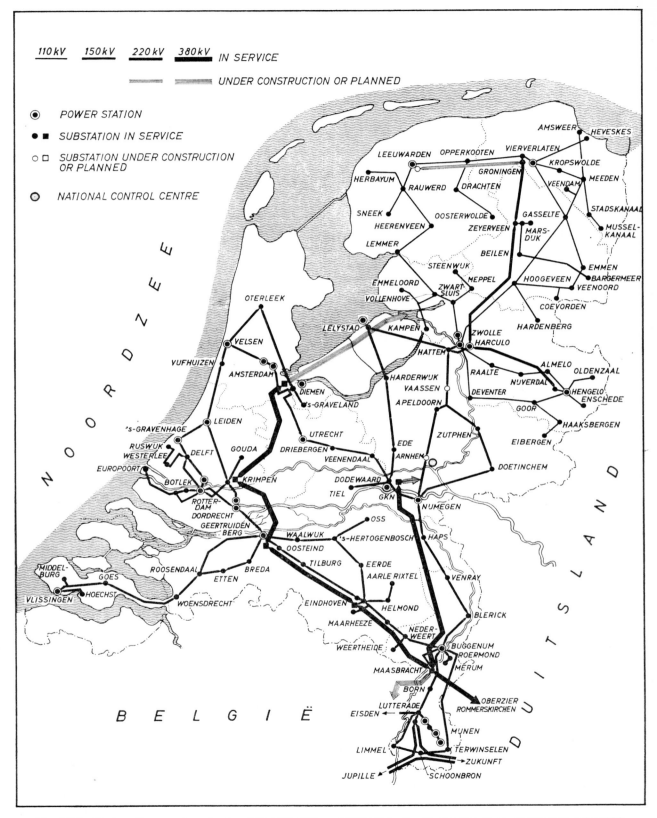

Fig. 73.5. High voltage networks for 380, 220, 150 and 110 kV in the Netherlands as on the 1st July 1970

After the introduction of the 380-kV network in the Netherlands the supervision of the partial networks for 150 kV will gradually become a task of the regional centres. These centres have already been executing switching manoeuvres in these networks, though not without permission from the national system control centre each time a line has to be taken out of service. The national system control centre supervises the 380-kV network in which it carries out all switching by means of remote control equipment. This centre further supervises the amount of generating plant in service as well as the allocation of the spinning reserve.

Frequency control is primarily the task of the regional centres under the supervision of the national centre, while the exchange of energy between the participating companies and with foreign countries is co-ordinated by the national centre.

The national system control centre at Arnhem, which was renewed recently, is equipped with a schematic diagram of the whole network which automatically indicates the switching situation. The centre also has telemetering facilities regarding production, energy exchange, loading of lines, etc. A computer arrangement was recently introduced and this will gradually be integrated into the security assessment of the system. However, much work remains to be done in this field, especially in connection with the software. As a logical consequence of the international collaboration within the UCPTE (i.e. the West European Union for the Co-ordination of Production and Transmission of Electrical Energy), SEP is working with its West European colleagues in the computer field to discover methods by which computer techniques may be used to aid the supervision of security in the interconnected system.

The economic load dispatch by means of digital computer techniques will be the subject of a national study. In that study special attention will be paid to the question of unit commitment. In the meantime tests are being carried out to establish the heat rate curves of all economically feasible production units on a uniform basis, and many data are being collected to get a better insight into the costs related to starting production units after different off-line periods.

THE LOCATION OF POWER STATIONS

Another characteristic of electricity production in the Netherlands is that, as a rule, sites for thermal power stations can be chosen such that cooling water can be drawn from rivers, estuaries, or canal systems. This is true for the existing situation, but in view of the very fast growth of electricity demand and of the need for cooling water in general, future possibilities should be carefully studied. Studies are being undertaken in close collaboration with the state water authorities.

The location of existing power stations or groups of power stations in the Netherlands is shown in Fig. 73.6. The size of the dots representing the power stations is

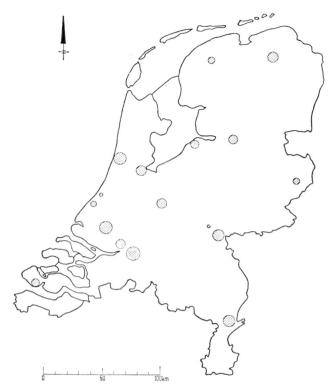

Fig. 73.6. Location and relative size of power stations in 1969

proportional to the capacity of these stations in 1969. There seems to be a reasonable dispersion of power stations over the country and this, compared with the spreading of load (shown in Fig. 73.7), does not show much need for bulk power transmission. Nevertheless, a strong interconnecting grid has been built in the Netherlands for the following reasons.

First, the interconnection of power stations allows for a considerable reduction of reserve plant as each station within the grid system can use the reserve power installed in the other stations. As a rule, a significant saving in investment costs remains after deducting the costs of such a grid. Second, in an emergency power stations can assist each other, thus avoiding a prolonged interruption of supply to a great number of consumers in the event of a severe disturbance.

The above also applied to the original grid for voltages of 150 and 110 kV, but by about 1965 the grid proved to be less and less capable of backing-up forced outages of big generators. There could be no increase in the size of new generating units unless drastic measures were taken to strengthen the existing grid. A new voltage level would need to be chosen. Noting the voltage levels used in neighbouring countries, levels of 220 and 380 kV were considered. The 220-kV level was rejected for the greater part of the country as the relief brought about by raising the voltage level from 150 to 220 kV would be of too short a duration unless a very densely meshed grid were built; and because of the prohibitive number of rights of way

Fig. 73.7. Spreading of load gravitation centres and relative size in 1969

that would be necessary, a densely meshed grid was out of the question.

Thus, for the greater part of the country 380 kV was selected as the new voltage level, provided that the investment costs of the superimposed grid might be expected to be offset by the economies attained by the use of larger thermal generating units, the choice of which was to be rendered possible by the introduction of such a strong grid.

A study was made at that time to estimate the economies that might arise from the use of larger thermal generating units in order to cover the increasing demand, including the necessary spare capacity. In this study only the savings in investment costs were taken into account. Further economies brought about by a reduction of the specific fuel consumption and by a decrease in specific costs of produced energy, associated with operating and maintaining a large thermal unit, were not taken into consideration.

Apart from the savings obtained from the use of larger units, there was the compelling reason that in the end it would be possible to meet the ever-growing demand only by increasing the unit size and not by a continual increase in the number of new units installed.

In order that the economies should not be overestimated, an investigation was made on the alternative assumptions of (a) doubling the maximum peak load in a 10-year period and (b) in a 15-year period. The result of

both calculations was that the economy to be attained from bigger generators was greater than the investment costs of a strong 380-kV grid.

CALCULATION OF FUTURE REQUIREMENTS

In 1969 generating capacity in the Netherlands amounted to 8510 MW, and this is to be extended in the near future.

The total generating capacity to be installed has been based on a probability calculation that was presented to a CIGRÉ conference in 1939*. In this calculation—based on the estimate of the load, the forced outage rates of machines and boilers, and an accepted risk level at the time of peak demand—a number of simplifications had to be made. (With the aid of modern computers, these simplifications are no longer necessary.) This led to a revision of the original methods, and after some reflection the principle of the 'loss of capacity' method, on which the initial calculation was based, was maintained.

The effect that the age of a production unit has on the forced outage rate is also taken into account. A fixed forced outage rate of 5 per cent for boilers, 5 per cent for turbo-generators, and 10 per cent for production units, respectively, had previously been assumed. It has recently been proposed that the following forced outage rates should be allowed for in the probability calculation:

11 per cent for production units less than 2 years old;
 8 per cent for units between 2 and 8 years old;
 5 per cent for units between 8 and 16 years old;
 8 per cent for units more than 16 years old.

The influence of the forced outage rates on the results of such a calculation is shown in Fig. 73.8 for three hypothetical cases. If the generating plant mix consists of units of equal capacity, namely 1·25 per cent units in the first case, 2·5 per cent units in the second, and 5 per cent units in the third, then the figure shows the influence of the forced outage rate (chosen as a fixed value for all units in this simplified example) on the necessary reserve plant to be installed. Taking the curve for 2·5 per cent units as being more or less representative of the average real plant, which consists of older (smaller) and more modern (larger) units, a 2 per cent decrease in forced outage rate would allow a saving of 5 per cent in installed capacity, i.e. one big unit. This may illustrate the extent to which the electricity supply companies are interested in more efficient production units.

The necessary margin of installed capacity, related to the maximum peak load, can be calculated as follows.

Assume that the production plant mix consists of units that can be in either of the following situations:

disturbed: probability of this situation p, capacity 0,
available: probability of this situation $q = (1-p)$, capacity V_1.

* BAKKER, G. J. Th. and VAN STAVEREN, J. C. 'Le rapport entre la puissance installée et la charge maximale admissible d'une centrale et d'un groupe de centrales interconnectées', CIGRÉ Report, Session 1939 (No. 331).

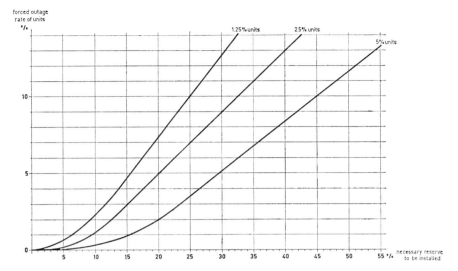

Fig. 73.8. Influence of forced outage rate on necessary reserve plant to be installed

Adding a second unit with a capacity of V_2 to the former, one can think of four possible situations. Adding a third unit, one can think of eight possible situations, by adding the capacities and multiplying the probabilities.

Continuing this procedure until all the generating units have been taken into account, one obtains a number of small intervals of load, in conjunction with the probability of being able to meet this load, which is the sum of the probabilities of the availability of the different production unit combinations, summing up to the capacity lying within this interval. From this there follows the probability that the available capacity will not be sufficient to meet loads which are higher than or equal to a certain value, which is called the probability of inability. From the probability of inability, as a function of the load, it is possible to derive the part of the installed capacity that has the prescribed accepted risk of unavailability. The difference between the installed capacity and this 'probably not available' part of the capacity should be equal to the estimated peak load. Thus, for a given peak load it is possible, via a number of iterations, to calculate the gross capacity to be installed for an accepted level of the risk of inability. The difference between the capacity to be installed and the estimated maximum peak load divided by this maximum peak load is called the necessary percentage margin. This margin will be approximately 0·28 in the near future. This relatively high value is connected with the rather low probability of inability accepted in the Netherlands, and also with the lack of hydro stations, which usually have low forced outage rates. Further, it is also influenced by the relatively large size of units. All new units will have a capacity of about 4–6 per cent of the total peak load of the country, which means that the total number of units is rather low. Nevertheless, it pays to have large units, as shown in the above-mentioned study where the savings from the use of large units, when compared with small units, were found to offset the extra invest-

ment costs in the new grid. The upper limit in unit size is, however, found by considering that this size should be related to the total demand of the area—equipped with centralized supervision of daily requirement—to ensure that the so-called contingencies can be met. In this respect the Netherlands relate their unit sizes to the maximum peak load of the country, and it is necessary to be able to overcome the forced outage of a large generator even in the situation where the country, for some reason, was not interconnected with the West European grid. As the ability of the boilers for immediate load pick-up is restricted to some 10 per cent of the capacity of the boiler, the loading of a big generator in summer should, as a rule, not exceed this value of 10 per cent. This agrees with a figure of about 6 per cent of winter peak load, which elucidates the limit set to the size of the largest units in the Netherlands.

THE ELECTRICITY PLAN

One of the main reasons for the decision to construct a super grid for the 380-kV system in the Netherlands was the willingness of the collaborating companies to realize that it was in the national interest to instal large generating plant even where these units were more than sufficient for local demand. This meant that the extension of the plant of the generating companies required a programme of closer co-ordination than had been experienced in the past. The electricity generating companies voluntarily agreed to establish such a programme; this programme is known as the Electricity Plan, and is reviewed annually.

In accordance with the rule explained previously, new units will be of a size corresponding to 4–6 per cent of the national highest peak load that is expected in the year in which the unit is to be commissioned. Exceptions will only be tolerated where and when they are well motivated. The foregoing rule means that, especially in the first years after commissioning, such units will be capable of generating more power than the amount that was required by the particular company when the new unit was considered.

After having estimated the national load and having taken the existing plant mix into account, the amount of new plant to be commissioned is determined by the use of the probability calculation. The electricity generating companies must then agree upon the sites on which new units are to be built and commissioned. According to agreement, the other companies that do not expand their generating plant in this year will have the right to import electricity from the new units built by other companies at a rate which corresponds to their shortage and at a price which is equal to the average of the production costs of the new units and which does not give any profit to the company owning the unit.

The constant increase in the demand for power causes the company that requested the extension of its power plant to need a steadily increasing part of the capacity of the new unit to cover its own requirements.

In the meantime a shortage of installed power plant may occur in other generating companies and when it is considered necessary in the national interest to construct a new generating unit, this will normally be assigned to the company with the greatest relative shortage. The other companies that have insufficient power in accordance with the probability calculation may draw electric energy from this company. It is possible that some of these 'importing' companies were supplying extra power only a few years before. It should also be stated that along with the right of neighbouring companies to draw electricity from new units at times of high demand there goes an obligation to ensure that such a unit can be loaded efficiently at other times, which means that they will have to import when demand is at a lower level.

This is part of the Electricity Plan, which covers a period of nine years. It contains a detailed programme of the first five-year period, which is obligatory for all participants. The intended programme for the following four-year period is based, by common agreement, on the estimated increase in demand.

One year later the plan is reviewed to cover a new five-year period and a new four-year 'intention' period, thus making the nine-year plan annually progressive.

Each Electricity Plan is named after the winter of the last year falling within the obligatory period. Thus in the recent Electricity Plan 1974–75, the last winter that fell within the obligatory period was that of December 1974 to June 1975, and the intention period in this plan covers the winters 1975–76 up to and including 1978–79. Extensions to generating plant in existing power stations, as well as the opening of new sites, are stated.

In the early years of the 1974–75 plan, the extensions will consist of units of about 400 MW, whereas 500- and 600-MW units will be introduced at a later date. By way of exception some smaller units are accepted where they are well motivated; for example, the 250-MW unit in Utrecht is the last unit that can possibly be constructed in the existing power station since the total installed capacity is limited by the amount of cooling water available. The two 300-MW units in Friesland will have to cope with the sudden increase in industrial power in the neighbouring province which has been announced at short notice; and owing to the short available erection time and the necessity to have fully reliable units as from the moment of commissioning in this case, the collaborating electricity generating companies are of the opinion that the choice of two relatively small units is better than, for instance, the choice of one 600-MW unit.

The electricity generating companies in the Netherlands also pay attention to the anticipated development of demand and of production equipment on a longer term than that covered by the Electricity Plan. In fact they study, within SEP, the necessary long-term development of the generating plant and the grid, and they ensure that the short-term planning, as shown by the Electricity Plan, will fit within this less detailed long-term planning. The choice of generating unit and fuel to be used also forms part of the considerations in the long-term planning, which includes the economies of constructing nuclear power stations. Until now, however, there has been no evidence that nuclear stations, contributing 4–6 per cent of the national power demand, are economically justified for the average load pattern in the Netherlands.

OPERATION OF AN INTERCONNECTED SYSTEM

The reliability and quality of electricity supply depend to a great extent on how quickly the power produced can be regulated to satisfy the amount requested by the consumers at any given moment. Since electric energy cannot be stored, it is evident that hydro power stations may be of great assistance in the regulation of power systems. Thus it will be understood that those countries which, from lack of hydro resources, have only thermal power stations at their disposal must be able to regulate the thermal units very quickly in order to assure a sufficient quality level. However, in the near future the need for rapidly regulating thermal units will not be exclusive to a small number of countries who do not possess hydro power stations, as the percentage of thermal production equipment is growing rapidly where the economically workable hydro resources have all been developed. This is also the case where large nuclear power stations are put into service, since here the economy dictates that they should be run continuously at a constant load level which, however, should leave sufficient latitude for primary regulation.

Thus, it might be of interest to learn the type of regulating properties that have been specified by systems which entirely depend upon thermal power stations. It should be remembered, however, that it might even be necessary to tighten up these requirements if, in the future, regulating duties were to be confined to only a part of the production equipment in service.

The daily load curves in the Netherlands show a

maximum rate of increase, on many days, of 1 per cent per minute. Assuming that the load variation during a continuous increase of load were divided over nearly all the units in service, then this increase in load would present no difficulties. In practice, however, one must also consider the possibility of a sudden outage of a generating unit in the interconnected system at any time.

Continuity of electricity supply can by no means be secured at all times. Savings have to be balanced against expenditure, and in this respect the Netherlands, as many other countries, have accepted the single contingency philosophy for the operation of generating units and of the high-voltage grid. This means that if the compulsory shutdown of a large generator occurs during the highest daily peak load, there should be no noticeable reduction of supply to the consumer; neither should the forced outage of one heavily loaded high-voltage circuit or one transformer have adverse consequences. A logical consequence of this method is that measures should be taken (a) to ensure that on the occasion of a forced outage of a generator, the primary regulation, the spinning reserve, and the following secondary regulation restore the balance between load and production and resume the proper frequency, and (b) to ensure that a single incident does not cause other apparatus to trip or to decrease the available capacity.

In the event of a forced outage of a big generator, the primary regulation via the turbine governors may call for a 6–10 per cent increase of generation within a time interval of approximately 30 s. This rate of change has to be added to that of the consumers' load, and it should be noted that, although the thermal conditions of the boiler may not yet have reached stable values at the end of this primary regulation period, the extra generated power should not fall off after this period, but should be maintained while all the regulating facilities of the boiler strive after a new stable position.

In that part of the network where the disturbance was caused, the secondary load frequency regulation will call for even more generated power as this partial network once more seeks to be self-supporting. This so-called secondary regulation may, however, take place at a slower speed, which gives rise to some additional though not very stringent requirements, provided that this regulation task is divided amongst a sufficient number of production units.

Based on the lines explained above one has come to the conclusion in the Netherlands that thermal production equipment, including the most economical units, should be able to pick up a 10 per cent increase in load within a time interval of 30 s.

It should be borne in mind that this requirement does not allow for the situation—which might arise, for example, some Monday morning—in which some units are brought on to the line rather late. Consequently, the boilers of these units would not start from stable thermal conditions when the load increase began, and this could lead to insufficient reaction from these units, resulting in a higher load pick-up stress on the other generating units. For this reason the requirements stated above should be considered to be minimum.

The power under primary regulation must be divided amongst as many production units as possible. The lack of quickly regulating hydro power stations in the Netherlands leaves the duty of primary regulation entirely to thermal production units, the boilers of which have only limited possibilities of load pick-up. For this reason too great a part of the spinning reserve should not be left to one production unit. It should be stressed here that a spinning reserve which cannot respond immediately is useless. This has been evidenced by some widespread blackouts which have occurred throughout the world in recent years. Consequently, to enable the boilers to respond quickly to frequency variations, one should restrict the load of all including the most economical generators to a level a few per cent below their continuous rating.

The latter statement leads to a careful examination of the turbine governor characteristics in the region of 95–100 per cent of the maximum load of the generators.

The statism of the turbine governor should not only be set to a rather low average value, but its characteristic should also be as linear as possible. Then the response of the turbine governor is optimal, irrespective of the loading of the generator. The value of this statism should preferably be the same for all production units as only then is a correct sharing of the regulating duties, which influence the stability of the steam condition of the boilers, obtained. The Dutch electricity generating companies have commonly agreed upon a value of the statism, which should be 4 per cent for new turbines and, where obtainable, the same for older turbines. However, the value for older turbines should not exceed 6 per cent. A linear frequency–response curve of the turbine governor can easily be obtained in modern electronic governors, but it has proved worth while to measure the linearity of the frequency–response curve of the older turbine governors. At one time there was some doubt in the Netherlands as to this item, as the response on frequency variations during disturbances—at that time—appeared to be different from what one would expect. Many older turbine governors appeared to have the characteristics shown, for example, in Fig. 73.9. Although the average value of the statism in this example (1) is 4·7 per cent, the response of the turbine governor in the region of 95–100 per cent of the rated capacity of the turbine is as poor as 22·8 per cent. It is in this very region that a turbine governor works when primary regulation is expected from the economic production units. It has been discovered, however, that it is quite possible to modify even the older turbine governors so that a more linear response curve is obtained. An example of such an improved curve (2) is also given in Fig. 73.9 (average, 4·3 per cent; 95–100 per cent range, 6·6 per cent).

The second consequence of the above-mentioned single contingency method is that a single incident must not cause other apparatus to trip or to decrease their available capacity. In other words, one should ensure that a single contingency remains single.

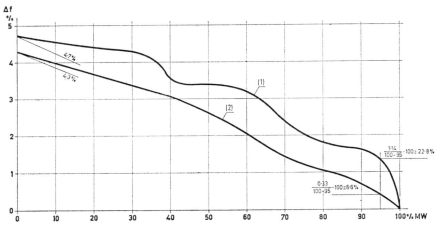

Fig. 73.9. Frequency–response curves of turbine governors

Thus lines and transformers of the high-voltage grid should be equipped with reliable protective equipment assuring the selective tripping of a faulty line or transformer. In the Netherlands the lines and transformers of the 380-kV grid have been equipped with double protective relays so as to assure a quick and selective elimination of faults, even when a protective relay fails to operate.

The above-mentioned rule also means, however, that a faulty production unit should be selectively tripped by its protective equipment and, further, that protective equipment of non-faulty units should remain stable during and after the occurrence of a fault. The greatest possible care should be taken in this respect to avoid unnecessary interweaving of supplies and auxiliaries of different production units which might cause simultaneous tripping of several production units on a trivial cause, such as a failure of the common supply of boiler flame protection. Neither should it be forgotten that production units may have many auxiliaries which directly or indirectly depend upon the voltage and the frequency of the main busbars. Both manufacturers and users of production units should constantly be aware of the fact that production units form an integral part of the electricity supply structure. The continuity of supply depends to a large extent on these production units remaining in operation and not tripping or diminishing their capacity during disturbances descending from elsewhere that cause abnormal frequency and/or abnormal voltages on the generator terminals and on the terminals of auxiliary apparatus.

In the Netherlands three extreme situations have been defined in order to make sure that production units, including their auxiliaries, deliver the service asked from them in any situation between normal service conditions and the extreme conditions specified. These extreme conditions are:

(a) a situation in which a frequency of 48·5 Hz and/or a voltage of 85 per cent of the nominal value exist for a maximum of 1 hour;

(b) a situation in which the frequency is 47·5 Hz and/or the voltage is 70 per cent of its nominal value for a maximum of 5 minutes;

(c) a situation in which the voltage of the main busbar is absent for a maximum of 3 seconds, the frequency remaining approximately 50 Hz.

The first extreme situation may arise as a result of a sudden shortage of production capacity after which automatic load shedding restores the frequency to 48·5 Hz and when, consequently, a heavy transport of active and reactive power brings the voltage down to 85 per cent of its nominal value in a restricted isolated area.

In this situation production units must be able to maintain their rated output. Production units which were out of service should also be able to be started up and synchronized under these conditions.

In the second extreme situation, production units which are in service must be able to remain on the line, if necessary with reduced capacity.

Under the third condition, production units must be able to be kept on the line for at least 3 seconds and resume production automatically when normal conditions return.

Another very important requirement is that production units that have been separated from the busbar (for instance, when the busbar voltage remains interrupted during a period longer than 3 seconds or as a result of a busbar fault) should maintain speed and maintain their auxiliaries under practically no-load conditions for a period of at least 1 hour, and preferably longer.

Only in those circumstances is a relatively rapid return to service possible after a severe fault, and it should be emphasized that failure to meet these requirements was at the root of some recent widespread and prolonged blackouts abroad. Further, experience has taught that while it is necessary to have tests to ensure that the requirements are adequately fulfilled when the unit is commissioned, it is also necessary to repeat site testing periodically to verify that the production plant still satisfies these requirements. This is the practice with modern protective equipment of lines and transformers.

Lastly, it should be stressed that, with every proper precaution being taken in the design stage, it is the task of the operation department to ensure that in everyday operation the possibilities for safeguarding the continuity of the electricity supply are used to their full extent. This means, for example, that the production unit with its auxiliaries should, in normal service, be as independent as possible from other production units; where emergency cross connections in auxiliary supplies with neighbouring production units exist, they should by all means be regarded as emergency supplies and not be taken into operation under normal conditions.

MANUFACTURE OF LARGE STEEL PRESSURE VESSELS FOR NUCLEAR POWER STATIONS

M. C. VAN VEEN*

This paper will be limited to those nuclear pressure vessels that have been (and are being) built by the author's company, viz. steel pressure vessels for water-cooled nuclear steam-supply systems (pressurized water reactors, boiling water reactors, and pressurized heavy water reactors).

First, data will be presented to illustrate the most important characteristics of these nuclear vessels. Some of these characteristics, such as design pressure and temperature, internal diameter and height, location and size of nozzle penetrations, are specified by the builder of the reactor system. Other data, such as the choice of the steel quality with mechanical characteristics, wall thicknesses, flange and nozzle dimensions, are generally the supplier's responsibility. Fabrication of nuclear pressure vessels, in general, not only involves the actual manufacturing operations, such as welding and machining, but also includes design and stress analysis, material selection, and quality control in the delivery contract.

INTRODUCTION

THE PURPOSE of this paper is to report on the present state of the art of manufacturing large steel reactor pressure vessels for nuclear power stations. The paper will be limited to those nuclear pressure vessels that have been (and are being) built by the author's company, viz. steel vessels for pressurized water reactors (P.W.R.), boiling water reactors (B.W.R.), and pressurized heavy water reactors (P.H.W.R.). It is considered to be beyond the scope of this paper to describe these different types of nuclear steam-supply systems. However, it is pertinent to recall that in a B.W.R. vessel, saturated steam is generated that goes directly to the turbine generator; whereas from a P.W.R. (or P.H.W.R.) vessel, hot water under high pressure flows to steam generators, where steam is produced in a secondary system.

The expected increase in size, weight, and wall thickness of nuclear pressure vessels that will result from the growth in unit size of nuclear steam-supply systems in the future will also be discussed. Up to a unit size of about 3000 MW no major limitations are presented by fabrication facilities and methods, and these very large vessels can be made to the same quality and safety standards as those existing today.

As shown in Figs 75.1 and 75.2, P.W.R. and B.W.R. pressure vessels have in common their general shape: a cylindrical portion with coolant inlet and outlet nozzles, a hemispherical bottom, and a hemispherical closure head that is connected to the vessel body by a bolted flange construction. Such a removable head is required to facilitate the installation of the reactor internals and to permit the periodic refuelling of the nuclear core. Therefore this closure head must be designed for easy removal. The vessels under consideration are generally made from low-alloy steels, and all surfaces in contact with the coolant are covered with a stainless-steel (or Cr–Ni–Fe alloy) cladding.

The major differences between P.W.R. and B.W.R. vessels are caused by the different operating conditions. A P.W.R. vessel has a design pressure of about 175 bar ($\simeq$ 2500 lb/in²) and a design temperature of about 350°C ($\simeq$ 650°F). The control rod penetrations are located in the closure head. The high power density in the reactor core gives a relatively small diameter (see Fig. 75.3), but the high pressure results in thick walls ranging from 200 mm ($\simeq$ 8 in) for 900-MW vessels to 250 mm ($\simeq$ 10 in) for 1500-MW vessels. Weights for the same power range vary from 320 tonne to about 500 tonne.

A B.W.R. vessel has generally a design pressure of 85 bar ($\simeq$ 1200 lb/in²) and a design temperature of about 300°C ($\simeq$ 575°F). The control rod penetrations are located in the bottom, as are provisions for the internal circulation pumps.

The lower power density gives larger core and vessel diameters (see Fig. 75.3), but the wall thicknesses remain smaller than those for P.W.R. vessels, ranging from 140

The MS. of this paper was received at the Institution on 20th October 1970 and accepted for publication on 12th November 1970. 33
* Managing Director, Rotterdam Dockyard Company, P.O. Box 913, Rotterdam, Netherlands.

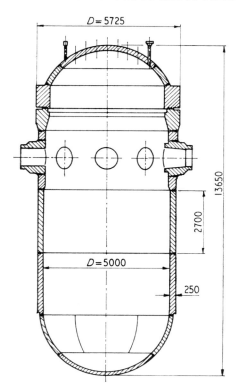

Fig. 75.1. Typical 1500-MW P.W.R. vessel

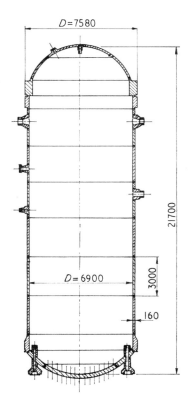

Fig. 75.2. Typical 1500-MW B.W.R. vessel

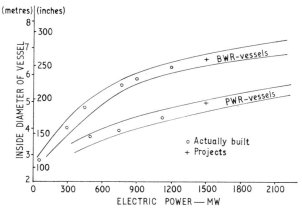

Fig. 75.3. Increase of vessel inside diameter
with electric power

mm ($\simeq 5\frac{1}{2}$ in) for 900-MW vessels to 175 mm ($\simeq 7$ in) for the 1500-MW type. Weights for this power range are 600 and 900 tonne.

The above-mentioned data—supplemented by other figures, such as the internal vessel height, the magnitude and frequency of thermal transients, the location and extent of mechanical loadings, etc.—result from the function to be performed by the reactor vessel within the nuclear steam-supply system. Therefore these data are specified by the reactor system builder to the manufacturer, who has the full contractual responsibility for ensuring that the design, materials, and workmanship shall be in accordance with the system specifications. The manufacturer's design responsibility includes such items as wall thicknesses, flange and nozzle configurations, resistance of the vessel against cyclic thermal and mechanical loadings, leak tightness of the seals, conformity with 'codes' and the regulations of regulatory bodies in the country where the vessel will be used, etc.

DESIGN AND STRESS ANALYSIS

The design of a nuclear vessel is determined by a functional (engineering) specification, which details the essential requirements in designing the vessel for its projected lifetime (normally 40 years), and also those items necessary to achieve integration in the overall nuclear steam-supply system, as mentioned in the introduction. In addition, the design must comply with the applicable 'pressure vessel code', which generally includes requirements from governmental safeguard agencies and establishes a 'code of practice', as agreed upon by the various experts in the field. In the last five years special codes for nuclear vessels have been developed, because the potential hazards are higher when compared to those connected with other pressure vessels, e.g. in the process industry. The additional hazards stem from the very high radioactivity contents of the reactor core. Examples of other differences which exist are the influence of neutron irradiation on the vessel wall and the problem of inspection after the vessel has become radioactive in service.

The boiler and pressure vessel code entitled *Nuclear*

vessels, A.S.M.E., section III, issued by the American Society of Mechanical Engineers, is perhaps the best known and most widely used code today. It includes requirements and guide-lines for design and stress analysis, mandatory and recommended requirements for manufacturing processes, such as welding and heat-treating, and it spells out in detail the minimum quality control requirements to ensure that a vessel built under the guide-lines of this code will be safe to operate during its designed lifetime.

The initial stage in the design of a pressure vessel is to establish the main wall thicknesses of the cylindrical and hemispherical portions. Here the code is a useful guide, because in it are listed the allowable 'design stress intensity values' for each of the code-acceptable materials. For the materials discussed in the next paragraph the allowable stress intensity at design temperature is about 19 kgf/mm² ($\simeq$ 27 000 lb/in²). Simple formulae are given to calculate these wall thicknesses. The task of sizing the flanges, nozzles, and other penetrations is much more difficult and requires experience from previous experimental work. A high degree of expertise on the part of the design engineer is necessary, because of the iterative nature of these calculations. If the vessel is to be built economically, and if it is to meet the code safety requirements, the ordering of the materials for its construction must be based on this experience and expertise. This demands a practical knowledge and complete understanding of the design under consideration, assisted by computerized calculations.

It has become standard practice to provide for each nuclear vessel a so-called stress report, which covers all static and transient pressure, temperature, and mechanical load conditions. In addition, the report includes a relevant appraisal of the static and fatigue loading of the vessel structure, using a diagram with stress classifications.

This stress report is based upon the vessel design made at an early stage of the manufacturing cycle, and is for use in checking whether or not the vessel will withstand all the different modes of operation during its lifetime. It is one of the aspects of nuclear vessel practice and procedure designed to prevent the possibility of discovering during the stress analysis work that the original design of the vessel will not cope with the requirements. For the stress analysis a set of computer programmes is required that will divide the construction elements of the vessel into small (finite) elements, by which thermal and stress distributions can be determined in often complex geometries.

Recently, the introduction of fracture mechanics methods into the design requirements for a reactor pressure vessel has been seriously considered. Fracture mechanics can be defined as an analytical method of predicting the stress load under which a crack or discontinuity of known size and location might be expected to propagate. A long-existing need is satisfied by the theory of fracture mechanics, to the extent that it becomes possible to correlate in a practical manner the allowable stresses in relation to the brittle fracture characteristics of a given steel and the inhomogeneities of the structure, sized and located by ultrasonics, radiography, or other non-destructive testing methods. Further discussion on design for fracture-safe performance is given in reference (1)[*].

MATERIALS SELECTION

At the present time the materials most commonly used for the pressure-containing parts of nuclear pressure vessels are A.S.T.M. A-508, class 2, for seamless ring forgings, flanges, plates, and nozzles; and A.S.T.M. A-533, grade B, class 1, for plates. The minimum tensile strength of this class of material is 56 kgf/mm² (80 000 lb/in²) at room temperature, the minimum yield point at room temperature is 35 kgf/mm² (50 000 lb/in²), and at design temperature about 28 kgf/mm² (40 000 lb/in²).

The chemical analysis for the above-mentioned steels (in wt %) is tabulated in Table 75.1. In the actual purchase specifications the carbon content is limited to 0·22 per cent maximum, phosphorus to 0·020 per cent maximum, and sulphur to 0·015 per cent maximum.

In order to obtain homogeneous material many precautions are taken. Not only is the P- and S-content further reduced, but it is also specified that the ingots should be made in basic electric furnaces and vacuum degassed. Grain refinement is obtained by adding some aluminium. Seamless ring forgings are often used where, during the forging process, the core of the ingot is removed. Should there be any segregations, this core contains them all.

Another important aspect is the necessity to obtain a high degree of toughness in the steel, as defined by the nil ductility temperature (N.D.T.) determined by the Pellini drop-weight test. This N.D.T. value should be −12°C (+10°F) or lower. To a certain extent it correlates with a Charpy V value of 5·2 kgf m/cm² (30 ft lb). Generally, both values are determined for reactor vessel forgings and plates. A frequency distribution was made of both Charpy V and N.D.T. values for 36 components for nuclear vessels manufactured in our company. Among these 36 components were shell ring forgings with thicknesses up to 315 mm ($\simeq$ 12½ in) and flange ring forgings with thicknesses up to 515 mm ($\simeq$ 20 in); the maximum ingot

References are given in Appendix 75.1.

Table 75.1. Chemical analysis

Material	A-508, class 2	A-533, grade B, class 1
C	<0·27	<0·25
Mn	0·5–0·9	1·15–1·50
Si	0·15–0·35	0·15–0·30
P	<0·25	<0·035
S	<0·025	<0·040
Cr	0·25–0·45	—
Ni	0·05–0·90	0·40–0·70
Mo	0·55–0·70	0·45–0·60
V	<0·05	—

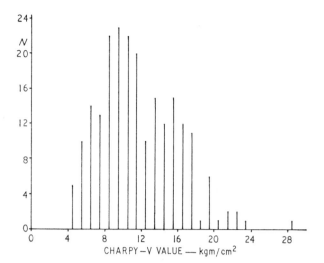

Fig. 75.4. Frequency distribution of Charpy V values at −12°C, individual bars

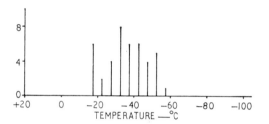

Fig. 75.5. Frequency distribution of N.D.T. values (A.S.T.M. E208)

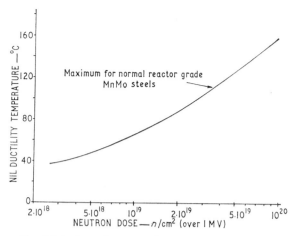

Fig. 75.6. Interference of neutron irradiation on N.D.T.

weight amounted to 160 tonne. All tests were carried out after a simulated post-weld heat treatment of 30 h at 605°C ($\simeq$ 1120°F) (see Figs 75.4 and 75.5).

To obtain the above-mentioned strength and ductility figures in the very thick walls and flanges of nuclear vessels it is essential that the standard normalization of these steels for smaller thicknesses be replaced by the so-called quenching and tempering heat treatment (rapid cooling from the normalizing or austenitizing temperature followed by tempering at a lower temperature).

The influence of neutron irradiation is of particular concern for those parts of the reactor vessel in the neighbourhood of the core. Generally, this is the middle part of the cylindrical portion of the vessel. The accepted variable in assessing the severity of neutron damage is the integrated flux of neutrons with energies above 1 MW. Depending on the reactor type and design, this integrated flux can go up to 10^{18}–5×10^{19} n/cm² during the lifetime of the reactor. The influence of most concern regarding this neutron irradiation is an upward shift of the N.D.T. values. Although this shift depends amongst other items on the chemical composition, the grain size, and the heat treatments of the steel, and can be different from ingot to ingot, a general trend is known. Many data have been established (2)–(4), and a general indication of the

maximum influence of neutron irradiation is presented in Fig. 75.6.

The interior of a water-cooled nuclear pressure vessel is clad with corrosion-resistant materials, such as stainless steels, or high-nickel alloys, such as Inconel. The environment is not actively corrosive but is still of concern because of the potential accumulation of radioactive corrosion products in regions of the coolant system where maintenance must be performed.

QUALITY ASSURANCE

The total effort taken to comply with the required inspection standards is generally referred to as 'quality assurance'. In a recent document (5) from the U.S. Atomic Energy Commission, *quality assurance* is defined as: 'comprising all those planned and systematic actions necessary to provide adequate confidence that a structure, system, or component will perform satisfactorily in service'.

Quality assurance includes *quality control*, which is defined as: 'comprising those actions related to the physical characteristics of a material, structure, component, or system that provide a means of controlling the quality to predetermined requirements'. These predetermined requirements in the quality assurance field are generally defined in a 'code of practice', as mentioned previously.

Further requirements are included in the equipment specification. The requirements included in these documents are not limited to pure physical characteristics but also contain quality assurance topics, such as the training and qualification of non-destructive testing personnel, production-record keeping, suggestions on company organization, etc. In addition to the quality assurance requirements imposed by codes, specifications, etc., it is assumed, as a matter of course, that the manufacturer will at all times use his own judgement and experience to assure the quality of the product he makes.

For specific manufacturing operations—such as cold-forming, hot-forming, forging, quench and tempering heat treatments, post-weld heat treatments, submerged arc welding, weld deposit cladding, manual welding, etc.—separate company procedures are established. Although the final result of these operations quite often will be checked later, when the operation under consideration is finished (e.g. radiography of the welds), proper quality assurance requires 'in process' inspection. That means that parameters, such as preheat temperatures, humidity of welding flux, voltage, and amperage of a welding process, etc., are controlled during the manufacturing operations. This is called 'in process' inspection, as opposed to 'after the fact' inspection, e.g. radiography (R.T.), ultrasonic examination (U.T.), mechanical tests (M.T.) (to check, for instance, the results of quench and tempering operations), dye penetrant inspection, etc. It is only by the continuous efforts of a manufacturing company to determine methods and procedures to improve 'in process' quality assurance that all 'after the fact' inspections can give good results. In addition to 'in process' inspection, other efforts should include personnel training, the use of proven

methods, repeated calibration of instruments and equipment used, procurement and receiving inspection of materials used, drawing and procedure change control, etc.

One particular aspect which should be given much attention is that of flaw detection. At present, all pressure-containing parts of a nuclear vessel are 100 per cent volumetric inspected by ultrasonics. This is not only done before the parts are inspected prior to assembly, but also, at least for the critical areas, after the hydrotest. A recurrent inspection of this nature is being considered during the lifetime of the vessel.

Surface-defect inspection is carried out by magnetic particle or liquid-penetrant methods. In addition, all welds are similarly inspected. Weldments are required to be inspected by radiography, and this presents problems for heavy wall thicknesses. Normal X-ray tubes are not sufficiently powerful for wall thicknesses over 4 in. For wall thicknesses from 8 to 20 in, only linear accelerators will

Fig. 75.7. Mandrel forging of a large ring for a B.W.R. vessel under a 3300-tonne forging press

produce the required quality of radiographs within a reasonable exposure time.

The manner in which a company will provide for outside requirements concerning its quality assurance performance will vary, depending upon the company's background, experience, nationality, and business relationships. In other words, there is no single solution to the set-up of a quality assurance programme.

This observation is one reason for the use of an individual company quality assurance (Q.A.) manual. Other reasons are:

(1) to formulate the company policy for quality assurance;

(2) to define clearly the quality system for company personnel, material suppliers, equipment purchasers, and inspection agencies;

(3) to assist company and departmental managers to check on the performance of the employees under their responsibility;

(4) to assist in personnel training and in outlining the job descriptions of all personnel directly or indirectly related to the quality assurance programme.

An important chapter of each Q.A. manual should be the *quality assurance plan*. For each contract involving a quality product a separate Q.A. plan should be made. This plan is made at an early stage of the project, and it should spell out all the inspections that are necessary before, during, and after each fabrication step. The Q.A. plan will indicate exactly when an inspection is necessary, who has to be present at that inspection, and at what stages of fabrication upgrading is required.

Upgrading is a complete verification as to whether or not a part of an order has been completely inspected, found acceptable (including eventual repairs), and is backed up by the necessary paperwork. The Q.A. plan also indicates which procedures must be used for each particular fabrication, heat treatment, and inspection operation. By making a sound and clear Q.A. plan, a large part of proper quality assurance is carried out.

The influence of modern quality assurance planning on

Fig. 75.8. Large welding positioner with B.W.R. bottom head

the manufacture of nuclear reactor pressure vessels is enormous. Perhaps this influence can be demonstrated in the most pregnant manner by stating that:

(1) The complete costs for all destructive and non-destructive tests, as well as for all other actions, that are necessary to assure compliance with the required inspection standards represent 8–12 per cent of the total costs of a nuclear pressure vessel, including all materials, delivered ex works.

(2) The total time involved in performing the actions mentioned under (1) is about five months, or 15 per cent of the total delivery time of a reactor vessel, which is 2½–3 years.

FABRICATION

As can be seen from Figs 75.1 and 75.2, the nuclear

pressure vessels under discussion are welded together from the following forged or formed components.

The main vessel and head flanges

These parts are generally manufactured as seamless forgings, or in exceptional cases they are welded together from forged slabs. The procedure for making large seamless rings starts with forging a solid cylindrical bloom from the as-cast ingot. The next step is to upset this bloom and pierce it by a hollow bar; consequently, the doughnut-shaped forging is mandrel-forged into the required size. As mentioned previously, during the piercing operation the core of the ingot is removed; this guarantees very clean metal because all eventual segregations are removed. Some companies have facilities to replace the mandrel-forging operation (Fig. 75.7) by a roll-forging operation. After forging, the flange is premachined, and then

Fig. 75.9. Stress-relieving furnace with large B.W.R. bottom head

quenching and tempering are carried out. Quenching is carried out in a large water tank, generally equipped with spray nozzles to permit accelerated cooling. After this heat treatment the destructive testing of test specimens taken from the actual workpiece is carried out. If the results of testing these specimens are acceptable, the final machining is carried out, followed by non-destructive testing.

The cylindrical portion of the vessel

This part is made from several rings. Depending on the diameter and wall thickness, these rings are made from bent plates with one or more longitudinal welding seams, or from seamless forging as described above for the flanges. In Europe all P.W.R. vessels are made from seamless forgings, originating from ingots up to 160 tonne, while B.W.R. vessels are generally made from plate. In the U.S.A. all nuclear vessels are made from plate. The advantages of forged rings are the improved cleanliness of the material and the fact that they are machined all over, which facilitates later fabricating operations. Plate bending can be carried out cold (under the tempering temperature) from already quenched and tempered material, or hot and heat treated after bending. The bending is done by rolls or

Fig. 75.10. 8-MV linear accelerator with top half of 800 MW P.W.R. vessel

presses. Some fabricators do weld the longitudinal weld by using the electro-slag method, and quench and temper the fabricated ring. After this, mechanical and non-destructive testing must be carried out.

The hemispherical heads

These heads are made from dished plates and, depending on size, from one plate or a central cap with orange-peel sections to complete the hemisphere. Dishing is performed cold or hot. After hot-dishing, the plate is quenched and tempered, and if necessary cold-sized. Then, of course, it must be mechanically and non-destructively tested. Plates can be obtained for these parts with sizes up to 4550 mm (180 in) in diameter, and up to 300 mm (12 in) thick.

The nozzles

Nozzles are always forged and, depending on size, hollow-forged or machined from solid bars. The machining of a nozzle into the required shape is a rather difficult job, because of the saddle-shaped welding preparation that fits into the cylindrical part of the vessel.

Joining the parts

The task of joining together the above-mentioned parts

Fig. 75.11. Large horizontal boring mill and milling machine

by welding is perhaps the most intriguing part of nuclear vessel fabrication. First of all, the technological aspect requires much development work in order to establish all the welding parameters, such as the welding process; the electrical variables; the choice of wire, flux, or electrodes; the preheat and post-weld heat-treatment temperatures, etc. Most welding is done on nuclear vessels using the single or multiple wire submerged arc process.

Equipment

A second area of interest is that concerning the necessary equipment. The welding of circumferential joints is carried out with a set-up of rollers and a welding-head positioner; the welding of joints in hemispherical heads is performed with large welding manipulators (Fig. 75.8); and the welding of nozzles achieved with a special 'Rotomatic' nozzle welder. During the welding, preheat is maintained at 150°C (300°F) minimum. This preheat temperature is not allowed to drop below this minimum, and the post-weld heat treatment has to be carried out without the workpiece being allowed to cool down. Therefore special consideration must be given to the transport and handling of the pieces to be heat-treated, and the furnaces must be

Fig. 75.12. Large vertical boring mill

located in, or directly adjacent to, the shop (Fig. 75.9).

The non-destructive testing of weldments includes U.T., M.T. and R.T., for which a linear accelerator is necessary (Fig. 75.10).

Most of the above observations are also valid for the weld-deposit cladding of nuclear vessels. The reasons for this stainless-steel cladding have been mentioned earlier in this paper. At present, most cladding of this type is carried out by a submerged arc process, using either strip or wire electrodes.

One difference in the manufacture of nuclear vessels as compared with heavy vessels for the process industry is the large amount of close-tolerance machining a nuclear vessel requires. The flange connection requires accurate machining of bolt holes and the sealing surfaces. The control rod penetrations and core-support surfaces need to be in line with the intricate nuclear core and control rods that will be placed in the vessel. For this machining—which is carried out on large parts of the vessel before joining and also on the completed vessel after final stress relief—very large machine tools are required. Examples are shown in Figs 75.11 and 75.12. The vertical boring mill of Fig. 75.12 has a turntable of 8 m (315 in), a maximum height of the workpiece of $7\frac{1}{2}$ m (295 in), and a maximum weight of the workpiece of 400 tonne.

Because of the complexity of operations the total time span to fabricate a 1200-MW nuclear vessel from contract to delivery, including design and material procurement, will take 34–38 months. It is not always possible to fabricate a vessel within the shortest critical path time of the schedule, because adequate loading of all equipment in a pressure vessel shop will require many vessels to be manufactured simultaneously, and thus waiting times cannot be excluded. The assistance of critical-path and capacity-loading computerized planning systems are essential for proper control of the flow of work in such a shop. The necessity for the most economic use possible of a pressure vessel shop is amplified by reason of the tremendous

Fig. 75.13. General view of 750-tonne nuclear welding, machining, and assembly bay

Fig. 75.14. Hydrotest of a P.W.R. vessel

capital investment that is represented in the equipment described above.

At the author's company a new bay was added to the existing pressure vessel facilities in the first half-year of 1970 (Fig. 75.13). This bay and the existing shops contain all the equipment described above. The heavy crane of 750 tonne that is available (with the possibility of adding a second crane of the same size), the dimensions of this bay, and the installed equipment are designed to make the fabrication of the largest nuclear vessels possible up to 3000 MW. The height of $26\frac{1}{2}$ m (85 ft) under the crane hook of this bay is necessary to permit the vessels to be

brought into a vertical position for the hydrotest (see Fig. 75.14). This test is carried out with demineralized water at a temperature of 35 degC (60 degF) above the maximum N.D.T. value determined for any of the vessel parts.

CONCLUSION

An attempt has been made to describe in a general way the problems and possibilities of nuclear vessel manufacture. The fabrication of these large vessels has now left the experimental stage and can be considered, at least for those companies that started early enough to enter this field, as a settled but highly sophisticated and specialized industrial venture.

It has been shown that no essential limitations are present that should prevent the manufacture of nuclear vessels for the largest electricity generating units that can be conceived today or in the near future. Although more work is to be carried out on the understanding of brittle fracture, nuclear vessels can be made with a very high degree of reliability and safety.

APPENDIX 75.1

REFERENCES

(1) PELLINI, W. S. N.R.L. Rept 6957, 1969 (Naval Research Lab., Washington).
(2) CARPENTER, G. F. et al. Nucl. Sci. Engng 1964 **19**, 18.
(3) STEELE, L. E. et al. Nucl. Applic. 1968 **4**, 230.
(4) BRANDT, F. A. and ALEXANDER, A. J. A.S.T.M. S.T.P. 341 1963 **9**, 212.
(5) UNITED STATES ATOMIC ENERGY COMMISSION release. 'A.E.C. provides additional guidance on quality assurance for nuclear plants', 1969 (17th May).

COMBINED CYCLES:
A GENERAL REVIEW OF ACHIEVEMENTS

B. WOOD*

The combination of a gas turbine with steam plant has been carried out, first, in the Velox boiler and, later, in two different ways described as the 'high-efficiency combined cycle' and the 'recuperation cycle' respectively. The first can in principle be applied to any steam plant irrespective of steam conditions and shows a gain of about 5 per cent on heat rate over the highest steam plant efficiencies. The gas turbine, if properly matched to the boiler, is usually 15–20 per cent of the steam turbine output. Gas is the preferred fuel for the gas turbine, but the boiler may burn any fuel. The plant is best suited to high load factor duties.

In the pure recuperation cycle a waste-heat boiler produces steam from the heat in the gas turbine exhaust. The condensing steam turbine is limited to low steam conditions and about 50 per cent of the gas turbine output. Efficiency is markedly improved thereby. Supplementary fuel burnt in the exhaust can raise steam conditions and increase steam output, but at the expense of falling cycle efficiency and increasing complexity of the boiler. Gas is the usual fuel, but any clean fuel is acceptable if low enough in price. Plant cost is moderate and performance suited to middle load factor duties.

The relationship between the two cycles and the question of transition between them is discussed by the aid of a new diagram. Minor variations are possible within each type, but a clear distinction is to be made between the two basic types, especially in capital cost.

HISTORICAL OUTLINE

THE IDEA of combining a gas turbine and steam cycle is old. It was applied very elegantly in the Velox boiler of 1932 (**1**)† where a compressor was put before the combustion chamber and a gas turbine after it. The pressure ratio was low, 2·5 to 3, the object being not to improve efficiency but to cheapen the boiler by reducing its dimensions and furnish a packaged unit pretested at works. The industrial gas turbine grew out of this application, although it was then treated as a boiler auxiliary operating at variable speed to suit combustion air requirements and at a moderate temperature below 1000°F. The turbine output required was only about sufficient to drive the compressor, no particular credit being at first claimed for surplus net output.

Because the air was heated in compression, the usual air heater extracting heat from the stack gases could not be used. Instead, it was necessary to revert to a low-temperature economizer, which in turn displaced the bled-steam feed heaters.

The Velox boiler was never built in larger sizes than 75 ton/h or for high pressure or highly efficient steam

The MS. of this paper was received at the Institution on 14th September 1970 and accepted for publication on 12th November 1970. 24
* *Associate and Technical Adviser, Merz and McLellan, Milburn, Esher, Surrey.*
† *References are given in Appendix 76.1.*

cycles, and it is no longer made by Brown Boveri or its imitators, Sural in France and Foster-Wheeler (U.S.A.). It is, however, being resuscitated currently in the Steag plant referred to in Table 76.1 in the combination of a highly pressurized gas-fired boiler associated with a Lurgi gasifier for the complete gasification of coal without oxygen.

A basically simpler idea was to put a suitable, complete, separately fired gas turbine in front of an existing boiler and so provide the boiler with hot combustion air, albeit depleted to about 17 per cent O_2 from the initial 21 per cent, and thus save the gas turbine stack loss. This proposal was made by the author in 1939 for Dunston 'A', but the project then showed insufficient return and was not proceeded with.

By 1950, when gas turbines had become a commonplace employing at least notionally much higher initial temperatures, the prospects of gains in cycle efficiency by this combination were discussed (**2**). The erroneous view was then held by the writer that the prospective gain was seriously restricted by the necessity of dispensing with the low-temperature, bled-steam feed heaters. Some time between 1950 and 1960, when various combined cycles were reviewed by Seippel and Bereuter (**3**), the trick had been devised of diverting only some one-third of the feed water through the economizer. This meant that only a third of the gain by bled-steam feed heating need be

Table 76.1. Some large combined cycle plants

Installation*	Year in service	Fuel†	Gas turbine, MW	Steam turbine, MW	Efficiency on L.C.V.	
Type 1: High efficiency						
Horseshoe Lake, Oklahoma .	1963	Gas	25	193	39·7 / 39·8	T / Y
Apache, Arizona . . .	1964	Gas	11·3	81·6	38·5 / 35·4	G / Y
Riverton, Kansas . . .	1964	Gas	12·5	37·5	—	—
Hohe Wand, Austria . . .	1965	Gas and coal	12	67·7	43·7 / ~40	T / Y
San Angelo, Texas . . .	1966	Gas	25	85	40·7 / 39·2	T / Y
Nevinomyssk, Russia . . .	1969	Gas	37	163	—	—
Vitry sur Seine, France (2 units) .	1970	Gas and coal	42	250	39·6	G
Shikoku, Japan	1970 ?	Gas	34	195	—	—
Altbach, Germany . . .	1971	Gas	50	200	38·2	G
Essen, Germany (Steag) (coal gasification) . .	1971	Coal	74	96	—	—
Gersteinwerk, Germany . .	1974	Gas	50	350	—	—
Emden, Germany . . .	1974	Gas	50	400	>43	G
Type 2: Recuperation						
Korneuburg, Austria . . .	1960	Gas	2 × 25	25‡	32·6	T
Neuchatel, Switzerland . .	1967	Oil	19·1	7·1	28·9	T
Liege, Belgium	1968	Gas or oil	23·1	23·2‡ / 10·9	29·7 / 32	T / T
Dow Chem., Texas . . .	1968	Gas	43	20	41	E

Y = year.   T = test.   G = guaranteed.   E = estimate.

* There are numerous smaller plants not listed.
† The gas turbine burns gas or oil; the boiler may burn coal.
‡ Indicates with supplementary fuel.

sacrificed. The result was the 'high-efficiency cycle', which shows a gain of some 5 per cent in heat rate over the best steam cycles at somewhat reduced capital cost.

The gas turbine is ordinarily limited to some 20 per cent of the steam turbine output. Putting the gas turbine before the boiler has the advantage of retaining standard items and separating the gas turbine, boiler, and steam turbine contracts and responsibilities. It also removes restriction on choice of fuel for the boiler as well as on its size and steam conditions.

For many years the classic means of improving the poor efficiency of the gas turbine was the use of a regenerator. This is no longer fashionable. It saves fuel but does not increase the output on a given frame. Indeed, it tends to lessen it since it requires for optimization a lower pressure ratio than that for maximum output.

Attention accordingly has switched to improving efficiency and obtaining larger output without interfering with the gas turbine by installing in its exhaust duct a waste-heat boiler supplying steam for use in a low-grade condensing steam cycle. The output of the steam turbine can readily be about 50 per cent of that of the gas turbine, whereby the overall efficiency is improved in the same ratio. If supplementary fuel is burnt before the boiler the steam output can be somewhat increased, but at the expense of falling efficiency.

It will be understood that where circumstances permit, combination with other heat demands is also possible, e.g.

steam may be supplied to a process or gas turbine exhaust heat used to heat the feed to an existing boiler, as was done at Belle Isle in 1946 and again at Dow Chemical in 1968. Such arrangements may be called combined cycles but are here not discussed, the term 'recuperation cycle' being confined to self-sufficient plant for pure power generation.

The present review of achievements and prospects is confined to the high-efficiency cycle on the one hand and the recuperation cycle on the other, since these are the two generic types which have been installed recently in large sizes. The question of transition from one to the other is considered.

HIGH-EFFICIENCY COMBINED CYCLE

The combination of a gas turbine exhausting into a boiler and providing that with just the right amount of oxygen for combustion is here called the 'high-efficiency cycle' because it can show a better efficiency than the steam cycle alone. When applied to a steam cycle with high conditions, e.g. 2000 lb/in², 1000/1000°F, the gain can be of the order of 5 per cent in heat rate. When applied to poorer steam cycles the gain is larger. Optimizing for somewhat lower capital cost rather than good heat rate is also possible (4) (5). Since the installed cost per kilowatt of gas turbines is ordinarily at least some 25 per cent below that of large steam plant, there is a potential reduction of capital cost per combined kilowatt, say 5 per cent. This is a significant claim.

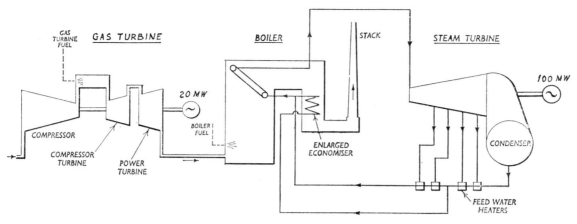

Fig. 76.1. High-efficiency combined cycle (basic concept)

Hitherto, improvements in steam cycles have usually had to be bought at the expense of higher capital cost. There is, however, a limitation; namely, that the gas turbine at least requires a clean fuel, generally natural gas. Classic examples of this type are Horseshoe Lake (6), San Angelo (7) in the U.S.A., and Hohe Wand (8) in Austria (see Table 76.1, upper part).

Horseshoe Lake cost $22 million in 1963, which on the rating attributed by the Federal Power Commission of 247 MW implies $89 per kilowatt. San Angelo, according to the same source (9), cost $11 million for 133·5 MW, making $82·2 per kilowatt in 1966. Vitry 2 (4), which saves only 1·55 per cent on heat rate, is claimed to save 10 per cent on the cost of the boiler. Overall cost is not quoted but is said to be low. Altbach, designed for peak load, is quoted as costing 280 DM per kilowatt (£29·5) (5). Figures in many cases are not disclosed.

Fig. 76.1 shows the basic form of circuit. It will be clear that because the boiler is now supplied with hot gases at, say, 700–900°F (depending on the gas turbine), the usual air heater exchanging heat with the stack gases cannot be used. Instead, the gases have to be cooled to stack temperature by a one-third flow, low-temperature economizer. This is additional to the full-flow, high-temperature economizer that would have been used in any case to bring the gases down to a suitable temperature at entry to the air heater. Because only about one-third of the feed needs to be used to cool the gases, the remaining two-thirds of the flow may still be heated by bled steam.

In the simplest plant the two circuits are in parallel. An alternative is to parallel the economizer only with the high-pressure (h.p.) heaters, leaving the low-pressure (l.p.) bleed heaters undisturbed as they are the most effective part of the process. Innumerable variations can be played on the arrangement of the feed circuit offering minor advantage in net gain, cost, or convenience according to the judgement or personal preference of the designer. An unusual variant is the use of a regenerator, as at Hohe Wand, which in combined working transfers heat from

the boiler-stack gases to the gas turbine after the compressor (Fig. 76.2).

The basic explanation for the gain in efficiency is that by transferring some fuel to the gas turbine, additional

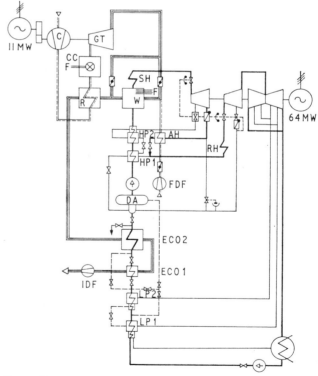

C	Compressor.	SH	Superheater.
GT	Gas turbine.	RH	Reheater
CC	Combustion chamber.	LP1, 2	L.P. feed heaters.
F	Fuel.	HP1, 2	H.P. feed heaters.
R	Regenerator.	DA	Deaerator.
IDF	Induced-draught fan.	ECO	Economizer.
FDF	Forced-draught fan.	W	Windbox of boiler.
AH	Air heater (steam).		

Fig. 76.2. Hohe Wand combined cycle

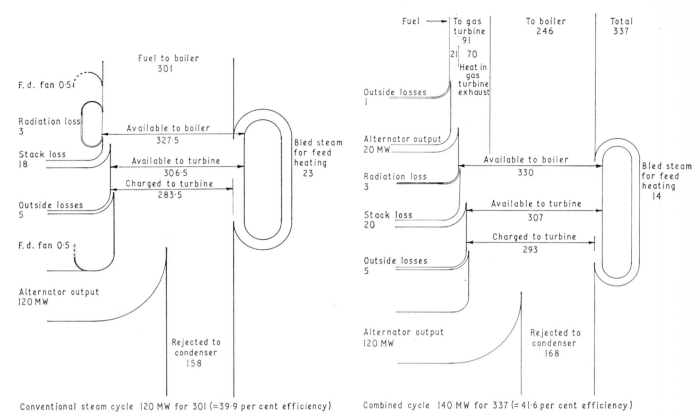

Fig. 76.3. Sankey diagrams comparing steam cycle and high-efficiency combined cycle

output is obtained greater than the sacrifice in the steam turbine and without increase of boiler stack loss, the total oxygen burnt remaining constant. The gas turbine so applied shows nearly 100 per cent efficiency in that nearly all of its heat intake is either converted to power or recuperated in the boiler.

The gas turbine cycle is to be regarded as superposed on the combustion air cycle. Hence, the thermodynamic gain is identical in principle to that obtained by superposing a high-pressure turbine in a steam cycle, except that extra output is then obtained without increase of heat rejected to the condenser.

The magnitude of the gain depends on the temperature employed in the gas turbine in relation to that in the steam turbine, and it can be assessed by considering that the credit to the boiler for heat in the gas turbine exhaust, say 800 degF above the stack temperature of 300°F, is some 11 per cent. This gain is offset by the sacrifice of one-third of the (10–12 per cent) gain ordinarily obtained by bled-steam feed heating, which amounts to, say, 4 per cent. A minor loss is attributable to the lower efficiency of the boiler when expressed on the 20 per cent diminished amount of fuel burnt in it for the same final oxygen content, without recovery of radiation losses but without fan debit (F.D.).

The Sankey diagram of Fig. 76.3 shows the heat balance. The overall result is a prospective net gain of some 6 per cent, which in practice produces $4\frac{1}{2}$–$5\frac{1}{2}$ per cent over steam plant with high steam conditions. There are many ways of arriving at the gain by short cuts, some of which give false answers.

Such gains are actually realized by the plants mentioned. For instance, Hohe Wand, with steam conditions of 2500 lb/in², 1000/1000°F, showed in acceptance tests on oil in the boiler and gas in the gas turbine an efficiency at 70 MW of 43·68 per cent. This is on net calorific value (net C.V.), as is usual in Europe. The efficiency in day-to-day operation is quoted as nearly 40 per cent.

Such figures are quite an achievement for a small unit since the best steam plant in the world in 1968 (the latest year for which statistics are available)—Marshall, with 2×386 MW at 2400 lb/in², 1050/1000°F—returned an annual efficiency of 39·3 per cent on high calorific value (H.C.V.) on coal, which implies 40·9 per cent on net C.V. The best Central Electricity Generating Board (C.E.G.B.) station, Ferrybridge C, with 4×500 MW at 2300 lb/in², 1050/1050°F, showed 35·23 per cent on H.C.V., again on coal, equivalent to 36·63 per cent on net C.V. Horseshoe Lake, with steam conditions of 1800 lb/in², 1000/1000°F, reported 36·3 per cent annual efficiency on H.C.V. in 1968, equivalent to 39·8 per cent on net C.V. This is about the same as its reported efficiency on test.

It is only fair to mention that in 1968 a new straight gas-fired steam station in the U.S.A., Robert Ritchie, with a 545-MW unit at 3500 lb/in², 1000/1000°F, did slightly better, and at a capital cost on F.P.C. rating of 640 MW was only $66 per kilowatt (the saving was largely accountable to size effect). Doubling the steam pressure at Horseshoe Lake should be worth about 4·3 per cent, which is about the gain attributed to the combined cycle.

It is rare to find statistical performances so closely reconciled with calculation. San Angelo, with steam conditions of 1450 lb/in², 1000/1000°F, showed 35·5 per cent in 1968 on gross C.V., equivalent to 39 per cent on net. Apache (10) shows an annual efficiency far below its expected 38·5 per cent, presumably due to occasional running on the gas turbine alone.

In the high-efficiency combined cycle the ratio of gas turbine output to steam turbine output is ideally fixed at something like 20 per cent. However, actual cases (see Fig. 76.7a later in the paper) show considerable scatter. Variations in gas turbine internal efficiency, initial temperature, and pressure ratio, partly also variation in steam conditions, account for some of this. However, the widest variation is caused by the choice of boiler rating at which the gas turbine is matched.

U.S.A. practice provides at least 15 per cent margin between name-plate rating and steam turbine capability. The boiler, and particularly its fans, accordingly have to be liberally designed to furnish this overload margin, which may be obtained at deteriorating efficiency. A gas turbine on the other hand swallows a mass of air fixed by its aspiration temperature and pressure. Hence, overload margin can ordinarily only be obtained by burning additional fuel, producing higher turbine inlet temperature.

If, therefore, the gas turbine is matched in mass flow to the steam turbine name-plate rating, the boiler's demand for additional air for overload can only be met by forced-draught fans used either in series or parallel. If on the other hand the gas turbine is matched to the full-capability rating, it will provide excess air at all lower ratings. Alternatively, the unwanted gases may be by-passed without benefit to the boiler. The choice is that of the designer.

Since it is usual to make provision for operating the boiler without the gas turbine in case of emergency, there must be a forced-draught fan to take the place of the gas turbine, ordinarily connected in parallel. Untypically at San Angelo, this forced-draught fan is placed in series with the gas turbine and thus boosts it. The duct work and the number of dampers are thereby reduced.

The more usual alternative involves dampers to control alternative supplies of air to the boiler wind boxes either from the gas turbine exhaust or from the forced-draught fan. When the gas turbine is not available the boiler may be supplied with cold air, or a steam-heated air heater may be used, as at Hohe Wand and in the Cesas system as adopted for the Vitry 2 plant of Électricité de France (Fig. 76.4) (Cesas = Chauffage de l'eau par surtirages de l'air par surtirages).

The gas turbine is also usually required to be able to

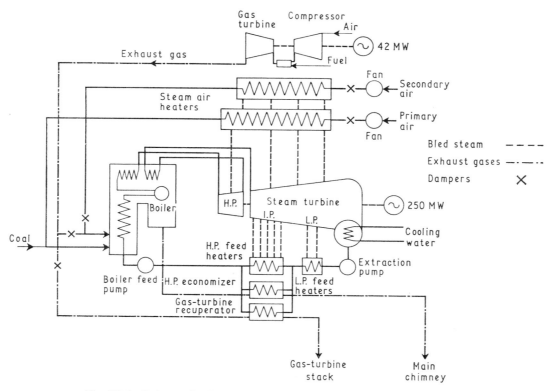

Fig. 76.4. Schematic diagram of Cesas cycle at Vitry power station

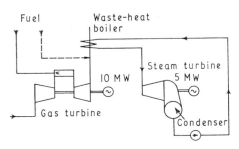

Fig. 76.5. Recuperation cycle (basic concept)

run alone. It then has to have its own exhaust stack. Gases can thereby be diverted round the boiler to any required extent. The boiler can also be heated up at a controlled rate before bringing on its firing. Dampers for these duties must be designed to withstand the gas turbine exhaust temperature and must be duplicated if work is to be permitted within the boiler combustion chamber while the gas turbine is running. Any danger in the converse condition is not significant. The boiler is substantially unaffected, except at the back end where the enlarged economizer replaces the usual air heater.

Steam conditions remain a free choice. Circulation may be natural or forced. A gas by-pass round the combustion chamber may be provided in addition to that via the chimney in order to improve part-load efficiency by picking up in the economizer some heat from the by-passed gases.

RECUPERATION CYCLES

In this process (in pure form) the gas turbine is the main power producer and some of its exhaust heat is recuperated in a steam cycle, as illustrated in Fig. 76.5. Choice of steam conditions is not free and they may have to be quite low, 250 lb/in², 600°F, in the example shown. The steam pressure is limited in accordance with the 'nip' temperature in the transfer of heat from the gases to the steam (see Figs 76.6a and 76.6b). The steam heat intake

line for a pound of steam (A, B, C, D in Fig. 76.6a) has to be fitted to the gas line EF, first by heat balance in that the heat from n lb of gas must equal that in 1 lb steam. According to the feed temperature at A and the desired stack temperature at F, the 'nip' may come at the gas-leaving end of the economizer F or at the gas-entry point opposite B. Superheat is also limited by condition D. Contraflow is assumed, which implies suitable differentiation of economizer, boiler, and superheater functions. A dual pressure cycle might be adopted producing greater output for a given 'nip' temperature (Fig. 76.6b).

Additional fuel can be burnt in the exhaust gases to superheat the steam additionally, to allow selection of a higher pressure, to increase the steam turbine output or cheapen the boiler. The boiler may thus for low conditions be a very simple unfired affair without combustion chamber, practically equivalent to an economizer, and may be designed to operate dry. It may even be shell type in small installations.

Alternatively, it becomes an elaborate water-tube type with a true combustion chamber if high steam conditions, obtained by supplementary firing, are adopted. A forced-draught fan may be provided to enable the boiler to run without the gas turbine. In that case, the plant evolves towards orthodoxy. In intermediate installations supplementary fuel, at least clean fuel such as gas, may be burned in the duct. Extended surface may be appropriate on clean fuels where no fouling is encountered. Fouling margin is then not called for. On natural gas containing little or no sulphur a stack temperature below 300°F might well be permissible. Circulation may be natural or forced.

With pure recuperation single pressure, the steam cycle may produce roughly half the output of the gas turbine; with additional fuel the outputs may be about equal. The efficiency of the recuperation case is easily derived; for instance, if the gas turbine efficiency is 26 per cent and 50 per cent extra output is obtained from the steam turbine, then the terminal efficiency of the combined cycle is 39 per cent. Any increase in output of the steam part by

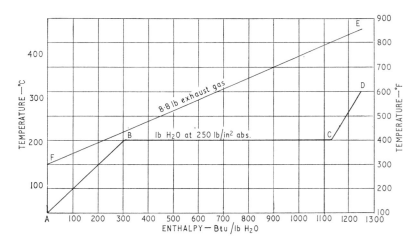

Fig. 76.6a. Temperature differences in recuperation

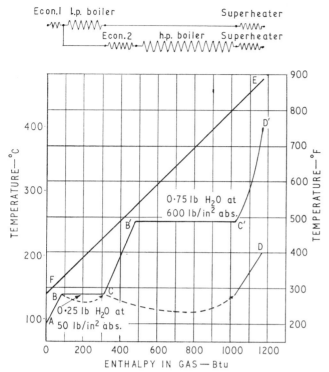

Fig. 76.6b. Temperature differences in recuperation—dual pressure cycle

burning additional fuel is obtained at low efficiency, namely that of the steam cycle *per se*, which may be no better than 20 per cent with a boiler efficiency of only, say, 70 per cent because of high excess air. Hence, the efficiency deteriorates as the proportion of steam output is increased.

In the above case 50 per cent extra output over the gas turbine at 26 per cent efficiency implies 13 per cent recovered from the 74 per cent heat in the gas turbine exhaust. Therefore the steam process efficiency is 17·6 per cent. This low figure is explained by the efficiency of the recuperation boiler being inherently very poor, since if the gases initially at 800°F go to the stack at 300, then the stack loss must be approximately 240/740, which is 32·3 per cent. Hence, the boiler efficiency cannot exceed 67·7 per cent. Thus in the example quoted the steam turbine evidently produces 26 per cent efficiency at terminals, which is about attainable with 200 lb/in², 600°F, if auxiliary power is disregarded.

Examples of this cycle are quoted in the lower part of Table 76.1, the largest being Korneuburg (11) (12), near Vienna, with 2×25-MW gas turbines associated with one 25-MW steam turbine running since 1959. Natural gas is the fuel and supplementary gas is burnt in part of the exhaust for the superheating of the steam, which is raised to 724°F at 200 lb/in². The tested efficiency was 32·6 per cent at 0°C (27·5 per cent for the gas turbine). The important characteristics of such recuperation plant are its simplicity and low capital cost, about £30 per kilo-

watt at Korneuburg. Although initially installed with the intention of running only 2000 h/annum, in fact it has run in the region of 7000 h and has proved highly satisfactory.

At Neuchatel (13) the overall capital cost per kilowatt of the 19-MW gas turbine with 7-MW condensing steam plant at £61 (1966) was about equal to that of the gas turbine—estimated efficiency is 28·9 per cent. There is no supplementary firing.

The Socolie (14), Liege, 23·1-MW gas turbine alone has an efficiency of 21·7 per cent. With recuperation of 10·9 MW, combined efficiency rises to 32 per cent on 34 MW. The steam output can be increased by supplementary firing to 23·2 MW, resulting in a diminished combined efficiency of 29·7 per cent on 46·3 MW (see Fig. 76.7).

The Dow Chemicals plant (15) in Texas produces 43 MW from a gas turbine and 20 MW from a back-pressure steam turbine on the same shaft. The gas turbine exhaust is used to heat the feed to a boiler, which with supplementary firing produces steam for the turbine at 1200 lb/in², 950°F. This exhausts at 185 lb/in² to process. The combined efficiency of 41 per cent here quoted is a notional one obtained by crediting 20 MW as the estimated output that could be produced by a condensing turbine operating at 235 lb/in², 700°F, and utilizing the gas turbine exhaust heat without supplementary firing.

It is to be noted that plant of this nature is not confined strictly to gas fuel. Neuchatel uses oil. The light and heavy oil-fired exhaust cycle discussed by the C.E.G.B. for Deptford (Fig. 76.8) is of this type and is unusual, first in coupling the power gas turbine to the main shaft, as at Dow Chem. (separate working is thus impossible), and secondly in the high steam conditions of 550 lb/in², 1000°F, with l.p. bled-steam feed heat. These conditions are attainable only by allowing a high stack temperature.

SELECTION OF GAS TURBINE

In every combined plant so far the gas turbine used has been of industrial type, generally of simple single-shaft construction. Exceptions (both industrial) are San Angelo, which has a split shaft, and Korneuburg, which has two-shaft reheat units. The selection has been influenced by the promoter being, in most cases, a maker of industrial gas turbines. It is of course desirable, particularly when a gas turbine is used in effect as an auxiliary to a much larger steam plant, that it should be extremely robust and reliable. When associated with base load plant it will be expected to run upwards of 8000 h/annum. Outage time for attention and overhaul is therefore to be a minimum, and this has been achieved.

The most satisfactory way of obtaining such reliability is to take a gas turbine of known good repute already used in base load service, or else suitably derated in temperature from established peak load duties. A low price can be obtained ordinarily only by employing a repetition article. Thus we are in effect driven to adopt a standard machine from both price and reliability considerations. This means that unless some standard gas turbine happens to fit the

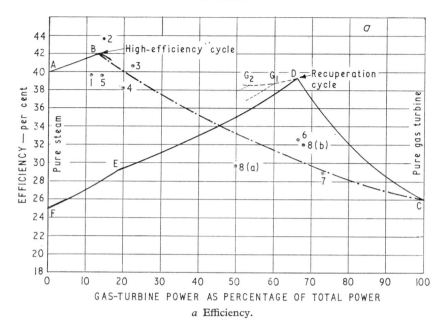

a Efficiency.

○   1 Horseshoe Lake. 2 Hohe Wand. 3 San Angelo. 4 Altbach. 5 Vitry.
6 Korneuburg. 7 Neuchatel. 8(*a*) Socolie with supplementary firing. 8(*b*) Socolie
without supplementary firing.

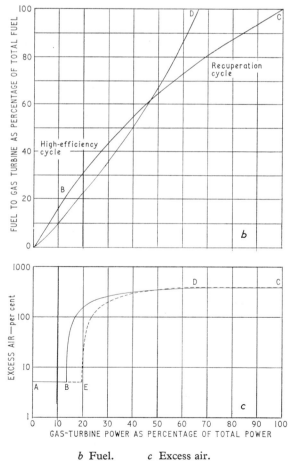

b Fuel.   *c* Excess air.

Fig. 76.7. Combined cycles—the gamut of possibilities

steam plant in mind, the steam turbine rating must be changed to suit the gas turbine. It is hardly practicable to adapt the other way.

Arrangements in which the gas turbine (or a power gas turbine) is mounted on the same shaft as the steam turbine clearly do not conform to the above principle, unless it is a combination package offered by the maker, and this arises only in smaller sizes. Furthermore, such single-shaft plant does not fulfil the usual desideratum of being able to use each part of the plant independently. A gas turbine used separately can be employed for emergency start-up or other emergencies, and thus can claim credit of capital cost for the deletion of the gas turbine that might otherwise have been provided for covering such emergency duties, including 'black start'. Industrial turbines are not normally expected to offer an especially quick start, 10–15 min being usual.

A jet aircraft gas turbine (not by-pass type) used as a gas generator in conjunction with a power gas turbine could fulfil certain of the above requirements, provided it were derated to obtain something like 10 000 h nominal running between overhauls. It appears, however, that when this is done the rating becomes so low, as well as the exhaust temperature, that the output for a given air quantity called for at the boiler is poor and the heat credit to the boiler is low. In the recuperation application the low exhaust temperature resulting from the high expansion ratio (except at peak duty) means the scope for power recovery is reduced.

Studies show that a simple gas turbine capable of a high entry temperature and of moderate efficiency and pressure ratio, and thus with a rather high exhaust temperature, seems to suit best. The use of variable speed is ordinarily

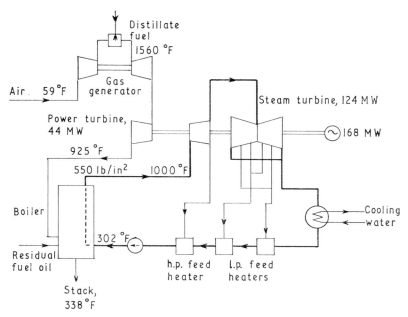

Fig. 76.8. Combined cycle discussed by the C.E.G.B. for Deptford

ruled out in that the power gas turbine, whether on the main shaft or driving its own alternator, has to run at constant speed. Two-shaft arrangements permit of some variation of air quantity, which might be welcome to the boiler since it requires close adjustment of air quantity to suit the fuel rate.

This can be achieved with constant-speed gas turbines only by by-passing the unwanted fraction of the constant air quantity. Whether this is done by putting all the excess gas to the stack or by-passing only the combustion chamber, so that the heat in the gases is utilized in the cooler parts of the boiler circuit, the efficiency of the plant falls away fairly fast at part load. The gain over orthodox plant accordingly falls to zero where the heat-rate lines cross at possibly half load. This means that the high-efficiency plant is best suited only to base load duties.

The part-load behaviour of a gas turbine with recuperation is much better than that of the gas turbine alone. The gas quantity produced by the gas turbine remains nearly constant at constant speed, irrespective of load. As load decreases, the stack temperature falls only slowly. Hence, the recuperation boiler can pick up if not a constant quantity of heat, at least one which diminishes more slowly than does the gas turbine load. Therefore the proportion of steam turbine output to gas turbine output increases as fuel is reduced.

If, of course, supplementary fuel is burnt to give additional output, clearly this is the first fuel to be cut off and hence efficiency rises on reducing load to the point where the gas turbine is on full load and supplementary fuel is zero. Thereafter it falls in the way described as the gas turbine load is reduced. Hence this type of plant has a fairly satisfactory part-load efficiency characteristic. Steam pressure may be varied also.

RELATIONSHIP OF THE TWO CYCLES

Fig. 76.7 shows a diagram devised to facilitate discussion of the relationship of the two generic forms of cycle. In Fig. 76.7a efficiency is plotted against the ratio of gas turbine output to total output. Thus on the left A represents a pure steam plant (gas turbine output zero) with moderately high steam conditions, say 2000 lb/in², 1000/1000°F, credited with 40 per cent efficiency on net C.V. (net C.V. is used throughout this discussion). Point B at about 13 per cent gas turbine output referred to total (gas turbine = 15 per cent of steam turbine) denotes a high-efficiency combined cycle showing 5 per cent improvement relative to A, namely 42 per cent combined cycle efficiency. (The optimum occurs at a ratio determined by the particular gas turbine, even though all gas turbines show effectively 100 per cent efficiency in this application.)

As we move to the right beyond B, having passed the point where the gas turbine flow exactly matches the boiler, the gas turbine becomes too large. The excess gases can at first be tolerated at the expense of increased stack loss but ultimately have to be by-passed, allowing the boiler to operate with no greater excess air than at B. If the gases are not directly by-passed but put through an economizer, this entails a further reduction of efficiency by omission of bled-steam feed heating.

The chain-dotted curve BC is calculated by combining appropriate proportions of the simple gas turbine as at C with the high-efficiency combined cycle as at B, in parallel. Existing high-efficiency plants are grouped near B. Other lower characteristics similar to ABC could be drawn for lower steam conditions, with the optimum shifting progressively somewhat to the right.

Starting from the other side of the diagram, point C

represents a gas turbine alone having an efficiency of 26 per cent. By recuperation of its exhaust heat in a low-grade steam cycle, say 250 lb/in², 600°F, we move progressively to optimum point D where steam output is about 50 per cent of gas turbine output, i.e. one-third of the total, efficiency 39 per cent (steam conditions being settled by the exhaust temperature and economics in accordance with Fig. 7.6a).

By the burning of supplementary fuel in the exhaust we can move still more to the left of D. This allows of some raising of superheat temperature and can, up to a point, be carried out with natural gas burnt in the duct without departure from the concept of a waste-heat boiler having no true combustion chamber. Conceptually, Korneuburg and Socolie are in the region to left of D and Neuchatel at D. Variations in assumptions shift the curves slightly so that the range of possibilities covers a band.

Assuming that we pursue this move to the left without change of steam conditions, we should ultimately reach point F where the whole of the output is obtained by the same low-grade steam cycle burning fuel directly with an efficiency of 25 per cent.

It is clear that when considerable supplementary fuel is burnt we are no longer limited to the low steam conditions imposed at D. Indeed, as the proportion of supplementary fuel is increased the cycle conditions can be raised progressively, leading to the dotted characteristics starting at points G_1, etc., and reaching successively higher intercepts on the Y-axis.

As the steam conditions are raised we ultimately encounter a discontinuity where the recuperation boiler has to undergo the change to an orthodox design with the superheater behind a convection surface, and with full burner and fan equipment (the discontinuity is not present in line DE).

Figs 76.7b and 76.7c show the relationships in terms of fuel burnt and excess air, which it will be seen is as low as $2\frac{1}{2}$–5 per cent from A to B and rises to a level of, say, 400 per cent from C to D.

A capital cost diagram might also be constructed. Capital cost is high along the characteristic A to B, and in its extension through to C remains high for the steam fraction because of the high steam conditions. Combined cost tapers because of the preponderance of gas turbine ultimately to gas turbine price at C. On the recuperation characteristics CDEF, capital cost from C to D remains low and almost constant per kilowatt of total output if we can accept the claim that the low-grade steam plant can be furnished at no greater cost per kilowatt than the gas turbine. In the region between D and B, capital costs change abruptly at the discontinuity from the low level to the high level.

Several studies over three decades have shown that on British prices no marked reduction of capital cost per kilowatt can be made for steam plant by choosing marginally lower steam conditions. The capital savings by adopting lower pressure, etc., are roughly offset by the

higher price of the boiler required to consume more fuel because of the lower cycle efficiency. Thus no plant in the transition region approaching B, where the boiler is no longer simple, is likely to offer low capital cost or particularly high efficiency.

ALTERNATIVE FLUIDS TO STEAM

The issue arises as to whether some other fluid, e.g. a refrigerant, cannot improve on the performance of steam in the recuperation cycle. Steam is ordinarily first choice because of its familiarity. Water is also cheap and abundant, and the cost and behaviour of plant is known with reasonable accuracy. Other fluids have the handicap that their precise physical properties are insufficiently known or at least not readily available.

Nevertheless, as illustrated in Fig. 76.6, steam suffers a handicap in taking in its heat in an inconvenient manner when viewed from the aspect of transferring heat from gases of roughly constant specific heat, implying an enthalpy–temperature relation in the form of a sloping line EF. The size of the heat exchanger is governed by the 'nip' or 'pinch point', wherever it occurs. This consideration limits the highest pressure that can conveniently be used and the mass flow of steam for any given waste-gas temperature when taken in conjunction with the feed temperature and the desired stack temperature. Any other fluid boiling below its critical temperature shows a similar heat intake line, though the ratio of latent heat to liquid specific heat is lower in almost all other fluids.

When used above critical pressure all fluids must be assumed to produce a smoother line. Since the efficiency of the cycle also tends to improve with higher pressure, there is a prima facie case for the notion that a fluid of low critical temperature could produce more power than subcritical water. Some fluids will avoid wetness loss, and many can offer a more compact plant by reason of their higher density. The drawback is, however, that the cost of the boiler increases fast with higher pressure. Some fluids are expensive and some are toxic or objectionable in other ways. Few complicated molecules are stable at temperatures as high as 600°F.

Hence, as gas turbine exhaust temperatures are almost always well above this level and are tending to rise, the instability of many refrigerants tends to rule them out. Finally, water can overcome its initial disadvantage by the use of dual pressure, as has been done in nuclear plants and as indicated in Fig. 76.6b (16).

FUTURE PROSPECTS

To assess the future we must first understand the present, in particular the apparent anomaly that the two combined plants in the U.S.A. installed several years ago, though successful, have not led to repeats there. Practical interest has shifted to other countries, while in England it does not seem to be able to get past the theoretical stage.

Two possible reasons can be seen for the standstill in

the U.S.A. First, for the last three years U.S. gas turbine makers have been too fully occupied with orders for peak and emergency gas turbines to be interested in other projects. The second is impending scarcity of natural gas consequent on its low price imposed by government control. This may seem paradoxical because some 75 per cent of the gas turbines being installed burn natural gas. The explanation is that a peak-load plant does not consume enough gas to be seriously incommoded if a higher price has to be paid, whereas a combined plant running on high load factor would.

Moreover, if gas is not available, peak plant can go over to distillate oil, the price of which would not preclude such utilization. The former moderate unit size is no longer a bar since industrial gas turbines have now reached 60–70 MW, which would fit steam turbines of 500–600 MW.

Furthermore, two gas turbines could be used (at some extra capital cost but with advantage at part load) to suit the largest units being installed. Several large combined units with 50-MW gas turbines have recently been ordered in Germany, one (Altbach) even for peak-load duty (5) (17)–(20).

England had until lately no natural gas, and partly as a result the industrial gas turbine has not developed in large units. With the advent of natural gas, the C.E.G.B., potentially the largest consumer, was barred from using it. This policy is in complete contrast to that in all other countries where on gas becoming available it has first been sold mainly to electric utilities. The reason being that they were the one consumer who could readily use large quantities, and so provide immediate revenue to cover the high expenditure on pipes.

However, the C.E.G.B. showed no great eagerness to use natural gas or appreciation of its advantages in a combined cycle, since E. S. Booth had averred (21) that only 1·9 per cent was to be gained. In coming back to the issue more recently, the C.E.G.B. face the prospect of base load soon being carried entirely by nuclear plant and hence feel little concern for the improvement of the efficiency of fuel-fired base load plant. Their interest has thus shifted to middle load factor duties.

CONCLUSIONS

Both types of plant have proved reliable in service and have been built to time and within budget price. According to the satisfactory experience in moderate-sized units, the high-efficiency combined plant burning natural gas in the gas turbine, and gas or any cheaper fuel in the boiler, appears able to offer various combinations of somewhat reduced capital cost compared with existing steam stations and proven higher efficiency.

This produces genuine base load overall generation costs lower than anything yet attained by nuclear plant. Some of the benefit is of course attributable, where gas is the only fuel, to the much lower cost of a pure gas-fired station. Nevertheless, major future application may be inhibited, where generation is co-ordinated, through nuclear plants monopolizing the base load.

The recuperation combined cycle appears capable of fulfilling a different role. Even discounting somewhat the claim that waste-heat steam plant costs no more per kilowatt than gas turbines, an economic mid-duty combination plant can be installed, at least in replanting an established site where circulating water works exist. With 39 per cent efficiency at the most economical point compared with 26 per cent of the pure gas turbine (said to be justified for 700 h/annum), the economic applicability of such low-cost combined plant on gas oil moves into the region of 1000–1500 h utilization per annum, whereas on natural gas even much higher utilization (4000 h/annum) may be profitable, as at Korneuburg.

Even 1000 h allows of running the plant regularly 12 h on every weekday in the four winter months and thus relieving the winter peak load on the distribution system in the middle of towns. The saving in this direction, of the order of £15 per kilowatt, appears sufficient to justify the higher cost per kilowatt resulting from the smaller combined unit size compared with a larger remote orthodox station. Amenity considerations favour such plant using clean fuel. For very low load factors, as in emergency and peaking service where rapid starting is the main requirement, the complication compared with a straight gas turbine is unlikely to be justified.

APPENDIX 76.1

REFERENCES

(1) MEYER, A. 'The Velox steam generator: a supercharged boiler', *J. Inst. Fuel* 1934 (August), **7**, 316.
(2) WOOD, B. *Fourth World Power Conf.* 1950, p. 2032.
(3) SEIPPEL, C. and BEREUTER, R. 'The theory of combined steam and gas turbine installations', *Brown Boveri Rev.* 1960 (December), **47** (No. 12), 783.
(4) COSAR, P., STEVENIN, M. and WIDMER, M. '(Vitry 2.) Le cycle CESAS à réchauffeur d'air par soutirages vapeur et à réchauffage d'eau pas les fumées', Paper C1–192, *World Power Conf.* 1968.
(5) BERNHARDT, S. 'Neckarwerke 250-MW (Altbach) combined cycle plant', *Mitt. VGB* 1970 (June), p. 153.
(6) GEORGE, T. H. '(Horseshoe Lake.) The world's first large combined cycle generating unit: how is it doing?', *I.E.E.E. Trans. on Power Apparatus and Systems* 1965 (December), **84** (P.A.S.) (No. 12), 1182.
(7) COX, A. R., HENSON, L. B. and JOHNSON, C. W. 'Operation of San Angelo Power Station combined steam and gas turbine cycle (formerly Lake Nasworthy)', *American Power Conf.* 1967, p. 401.
(8) GÖBEL, K. '"Hohe Wand" gas turbine/steam-turbine power station', *Siemens Rev.* 1966 (May), p. 262.
(9) FEDERAL POWER COMMISSION (U.S.A.). 'Steam-electric plant construction cost and annual production expenses, 1962 to 1968'.
(10) SYBERT, W. M. '(Apache.) A medium-size combined power station', *A.S.M.E. Paper* 66–GT/CMC–69, 1966.
(11) AUER, W. P. 'Practical examples of utilizing the waste heat of gas turbines in combined installations', *Brown Boveri Rev.* 1960 (December), p. 800 (covers Korneuburg and others).

(12) GRINDROD, J. 'Austrian gas turbine–steam cycle (Korneuburg)', *Engng Boil. House Rev.* 1963 (August), p. 290.

(13) ZÜND, L. 'The combined gas-and-steam turbine plant of Électricité Neuchâteloise SA', *Sulzer Tech. Rev.* 1967 (No. 4), p. 218.

(14) SCHÄDELI, R. 'The Socolie gas-steam station', *Sulzer Tech. Rev.* 1968 (No. 4), p. 159.

(15) GASKINS, R. C. and STEVENS, J. M. 'World's largest single-shaft gas turbine installation (Dow Chemical, Freeport, Texas plant)', A.S.M.E. Paper 70–GT–124, Brussels, 1970 (May).

(16) WOOD, B. 'Alternative fluids for power generation', *Proc. Instn. mech. Engrs* 1969–70 **184** (Part 1, No. 40), 713.

(17) 'Gasdampfturbinenkraftwerk mit Druckvergasungsanlage (System Steag)', *BWK* 1969 (December), p. 648.

(18) ROSAHL, O. 'Deutscher Dampfkesselbau', *Mitt. VGB* 1970 (December), p. 457.

(19) 'Emden Unit No. 4. 450-MW combined cycle', *BWK* 1970 (October) (No. 10), p. 501.

(20) ROSAHL, O. 'Gersteinwerk 400-MW combined cycle plant', *Mitt. VGB* 1970 (December), p. 456.

(21) BOOTH, E. S. *Sixth World Power Conf.* 1962, p. 4679.

PLANNING, DESIGN, AND CONSTRUCTION OF LARGE THERMAL POWER PLANTS

H. H. GOTT*

The author offers observations on the organization of work for a large modern power station, having regard to the interaction of technical, financial, labour management, and allied questions. The difficulty of integrating a number of specialized disciplines into a working organization is equalled by the need to draw together a number of separate organizations for a common objective. The difficulty of negotiating meaningful guarantees implies strict control of the amount of innovation introduced into a contract, while the risks appear to increase rapidly with the complexity of the contract. The subject is of general interest and of economic importance. It has had little treatment in papers to professional institutions. Though it is not discussed either so surely or rigorously as a more exact subject, its importance justifies considerable effort to achieve clarity on the major issues.

INTRODUCTION

THE MANAGEMENT of a large power station project differs in some respects from that for other large-scale capital projects, and has a distinctive character. This arises from the interdependence of the design content of the major 'blocks' of work, the amount of work involved in each of these blocks, the role of precedent work for similar stations, and the difficulty of developing and applying completely rational management techniques in a field of work which has many traditions and many conflicts of interest.

From the point of view of an operating utility it can be argued that to avoid building such advanced plants, but always to use the latest equipment that is fully understood, is good business. In this sense thermal power plant differs significantly from chemical process plant, where innovation is so often essential to provide the new products which are demanded by the market. For the electric utility, novelty can only be introduced against the expected economic justification without change of product.

The establishment of a new power station is, in general, a task arising from extrapolation of observed demand for electricity, and it is not our intention to discuss techniques of forecasting. For a nationalized utility such a plant is but one of many 'essential' competitors for Government allocation of funds, competing for attention against quite dissimilar schemes. For a privately operated utility, although the sources of funds may be different, market forces operate in deciding the allocation of funds and the power plant appears to investors as but one 'option' among many.

The plants have, correspondingly, to be 'sold' before they exist—sold to the source of funds on a description, backed usually by a cost benefit analysis.

We cannot, of course, say precisely how much each hypothetical plant will cost either in capital or in operation, though some, usually relatively small, influences can be predicted with near certainty. There is a tendency to forget that there is no sound tested theoretical basis for the comparison of quite different types of novel equipment, and the use of techniques, however well adapted they may be, for assessing the effect of small adjustments to operating plants has not, when applied to discriminate between complete plants, given many results which can be quoted as useful, though the technique has given some which are very disturbing in outcome.

Any discussion of the merits of various hypothetical plants is, and must be recognized to be, in its very essence, probabilistic. In this, of course, it differs in no way from all other human activities, but much misunderstanding would be eliminated, and management situations clarified, if more thought were directed to the understanding of this matter and to the assembly of meaningful data on the probable outcome of the various decisions involved.

MANAGEMENT OF THE PROJECT

The plant under discussion will be, perhaps, of 2000-MW capacity, and whether nuclear, oil, or coal fuelled, the

The MS. of this paper was received at the Institution on 19th January 1971 and accepted for publication on 29th January 1971.
* Managing Director, Associated Nuclear Services, Dorland House, 14–16 Regent Street, London S.W.1.

management tasks will contain many points in common. The project team and its leadership must have allocated to them the authority necessary to do the job. There is no complete agreement on the meaning to be given to this expression, and yet it is quite fundamental to regular and orderly working, and the writer finds discussion is often confused.

The task must be defined in the simplest possible terms that can be adequate, and the responsibility for organizing the work should be allocated to one man. At the same time this man must be given authority which is adequate to secure a discharge of his responsibility. If this is not done —and this is very often the case—it is nonsense to regard him as being truly responsible for the work, and such pretences should, among engineers, be rated as poor professional conduct. Among the sets and subsets of activities we may distinguish 'organizing', 'management', and 'technical', and it is not adequately recognized that each can have precedences over the others.

The utility, or in any case, the source of funds will usually give an 'authority to proceed'. This 'authority' is very nearly identical in standing with the 'approved budget' and it is at this point in time that the project and its organization are defined, however loosely this may be handled. In reality the 'project description' should include explicit statements of methods of working, of the financial authority, and of approved organization.

The data which are available on the success or otherwise of power plant projects are obscure and often unreliable. It is frequently actually misleading to ascribe success or failure to some particular characteristic of any one project, for the factors influencing the work are so many and varied. Consequently, one can only speak usefully in terms of probabilities, taking or recommending that course of action which seems to have the optimum chance of achieving the ends set out in the 'project description'.

When we start talking like this we are approaching the question of guarantees, and this must be discussed before questions of organization and management can be clarified.

GUARANTEES

The purchase of an article *consuming* electricity may be, and frequently is, covered by a guarantee which will, should it be invoked on account of failure to perform, leave the purchaser suffering negligible loss. No such situation can arise in relation to the purchase of large-scale plant for the production of electricity, for the earning capacity of such plants is sufficiently great that such a guarantee is quite beyond the means of any supplier. Moreover, the 'money back' guarantee concept could have only limited utility unless it were possible to produce 'instantly' the deficit in product output from the guarantee money.

A guarantee can ensure that the guarantor's attention is fully focused on the subject matter of the guarantee, and can operate in at least two distinct ways. The application of stringent penalty (or penalty/bonus which is closely similar) clauses should discourage a prudent tenderer from offering untested wares. Another effect of a guarantee is to assist the project executive in maintaining its 'rightful' position in all the queuing situations which the job will encounter. There is no single, simple discussion possible of guarantee questions, and any attempt at a proper treatment involves a good deal of speculation. Any attempt to get a contractor to guarantee something of which he does not have experience that is statistically significant is, at best, a gamble between the two parties.

Emphasis has shifted so that prompt and high availability are seen to have the same importance as thermal efficiency and output. Those aspects of a machine which determine its performance (especially its performance when new!) have a basis of experimentation and calculation which is very solid, and indeed, relatively simple. The techniques of analysis are simple enough to form the basis of part of the syllabus of heat engine courses in university and polytechnic. The financial losses due to inadequate performance of turbo-generators have, over the last few years, been very small, though the writer has known more difficulties with boiler surfaces wrongly estimated. Here the balance between secure, tested theory and rather *ad hoc* empiricism is less favourable.

The theoretical/empirical basis for discussion of probable outage is very much weaker. A major function of standardization of equipment is that it can significantly improve the rather poor statistical basis for assessment of probable outage, but before real progress is made there is need for a more genuine exchange of information than is at present attempted. Many tables of data on this subject are either valueless or actually misleading.

The best assurance of prompt and good availability in any completed plant will be secured by strict control of the innovative content of the plant items. This is not to be taken to imply avoidance of all change, but rather to insist that all changes made should be specifically in the interests of availability.

Such a debate leads to the conclusion that only what is fully understood can be guaranteed, and that to offer or accept a guarantee for anything that has not reached this state of development is to involve oneself in confusion.

Considering the need for a basis of fully analysed experience before success can be guaranteed, appears to lead to the view that (a) under some circumstances, plant items may usefully be the subject of guarantees, or (b) complete plants may only be the subject of rather ineffective guarantees.

The really useful area for discussion is (a), and this involves a knowledge of the statistical basis for assertions of probable failure rates. The range to be found in power station practice is very great, from electrical relays for instance—which can be thought of as fully understood from the knowledge of the behaviour of a large population, and in which the mode of failure under discussion is very simply described—to, say, a coal mill, the first of its kind, but with rather similar, usually smaller, ones which have been 'reasonably satisfactory', with, however, a completely different type of coal.

For the term 'guarantee' to have any clear meaning it must be in respect of a component or plant item, which is fully understood, to be operated within prescribed limits, and with recording instruments which beyond falsification make clear the way in which the equipment is actually operating.

If more than a very small number of closely linked items are linked together in a 'guarantee', then it is best to realize that one is buying 'goodwill' rather than a guarantee as such.

ORGANIZATION OF WORK

The 2000-MW plant which forms the background to this argument comprises many activities, and it requires a small team to organize them and secure a review of progress in such a manner as to ensure that the objectives are met. It is an important attribute of such a team, which is essentially one of leadership, that it should be confident that it will succeed. This is only partly secured by attention to personal qualities in selecting the three or four men who are to provide the 'drive', but also by careful attention to their terms of reference and to the forces available to support them should assistance be needed to overcome a difficulty, and especially in the reality or otherwise of the authority which this group exercises on the project and the extent of capacity of individuals or organizations outside this group, and not under its guidance and control, to interfere with, or even question, its decisions.

The purchaser's organization will have views on many features: appearance, general operational characteristics, quantity and standard of office accommodation, space and equipment for maintenance, among many.

Such views will be either requirements or wishes. If they are requirements they must be recognizable in the original project brief, otherwise they are at the option of the project manager.

The overall organization must be such that the project manager can himself settle all the compromises that are necessary. If he has to take instructions subsequent to his original brief, any incompatibility either weakens his authority or gives him an excuse for non-performance.

If the purchaser has separation between the operating function and that of designing and constructing the plant—which can either be in his own direct employ or may be an independent firm of engineers—then it must in any case be clear that the engineer acts in every way with the purchaser's authority and he must himself take and weigh all relevant information in taking his decisions. A most frequent cause of loss of control is the existence of confusion on this point. The project team must be absolutely clear as to its authority and whence it derives.

These are circumstances in which the purchaser decides to pass the control of the works to a contractor who will provide a so-called 'turnkey job'. Here the 'project brief' has overriding importance, and any amendments after the contract signing can be very costly.

Corresponding to the view that all the specialist functions must report to the project manager is the implication that he and his small group of leaders or supervisors must have a comprehensive understanding of all the relevant subject matter. At all times this is an absolute requirement concerning alike design, estimating and cost control, control of all interfaces, quality control, manufacturing progress and construction. The specialist advisers within the project team may have only part-time activity on its behalf, but any failure to resolve consequent priorities will jeopardize the work seriously.

Since the subject matter is broad and the discussion intensive, the control of paper work is important. In general, when paper is used as a medium for discussion, the job may be slowed down to a pace which makes for low productivity all round, but the correct formulation of schedules of acceptance records is essential. There should be only quite short-term reliance on individual memory. To this end it is important that contractual issues, estimates for variation in price, should be—must be—estimated for and agreed currently, or financial control will become a sham as relations also become strained. There can be no excuse for failure to arrange for this. A project which lacks financial clarity half-way through can hardly have a good outcome.

PROGRAMMING

The general technique of programming the work is now fairly well understood, though little of use has been published specifically for power station work.

From the foregoing discussion it is possible to modify slightly the traditional approach.

We assume the station particulars given and also the operational data.

The master programme describes the main strategy to secure this and the subsidiary programmes which must be consistent with each other are, in effect, the tactical positions which must be taken to carry out this strategy. If the totality of programmes is satisfactory it must be possible to go into sufficient detail on any subnetwork to verify the end dates, or at any rate to assign a probability curve to the dates involved. When we discuss programming we have partially left the relatively secure area of technical discussion and are involved not merely in management but in business. For the totality of contracts does not amount in any way to an agreement by the contractors to work together with the constructor to ensure success unless the contracts are deliberately drawn up to ensure this, which could be quite imprudent.

If, in a contract, it is possible for an excuse to be offered for non-fulfilment to time, then a programme based on such contract is to be considered quantitatively as being as insecure as the probability of the excuse being offered—which is, of course, higher than the probability of its being accepted.

One of the more difficult topics when one comes to discuss an overall programme is the allowance, or contingency, that is needed to ensure that a given 'end date' will be met. The writer's experience and impression of such discussions is not satisfactory. To speak of absolute security in a programme is of course nonsense, but we can have an acceptable probability of success. A good deal of data accumulation and processing is needed for a proper, quantitative discussion, but it is suggested that there should be no real difficulty in assigning acceptance criteria—they should be consistent with, and largely derive from, the calculations used in planning the system. When a number of units are under construction at one time, it is not at all obvious that one should aim at equal probabilities of completion to fixed programmes.

There has perhaps been rather too strong a tendency to transfer information from one programme to another without adequate analysis. For example, one is familiar with the notion that 'in the U.K. the best achievement in boiler erection suggests 24 months from the time of lifting the drum to the time of floating safety valves'. The more useful formulation might be: 'What measures can be taken to secure that this is achieved in 18 months?'

Whatever programme is ultimately adopted each 'event' must be scrutinized carefully as to the acceptability of its probability of success. If this probability does not appear high enough, then action must be taken to make it acceptable. This commonly implies making available additional resources in appropriate form. Every event must be solid enough to be relied on, and resources appropriate to this must be available to the project team, and the programme must be 'audited' with this in mind. In particular, the description of the event must be adequate and the activities on which it depends must be properly foreseen. In particular, attention must be paid to ensuring that all design work related to plant components —turbine, boilers, feed pumps, etc.—is (a) capable of being carried out so as to ensure satisfactory release of information for manufacture according to the programme, and (b) capable of being carried out in a manner that will allow of satisfactory information flow for all interface problems to be handled properly.

This analysis, or design programme audit, must examine the extent to which this design work is dependent on test and development work. The acceptability of such a programme is a very nice exercise in judgement, in which one is all too conscious of the tendency to programme on the assumption or belief that every test will be carried out to time with confirmatory results. It is true that there are such tests, and sometimes they are simply unnecessary. But it is vital that test work to support design is largely completed before the decision is taken to put the plant item into a power station on which the utility immediately is to depend.

Close consideration of this point might well lead to a slight but useful change in plant development technique. This would probably entail a separate treatment for plants whose overriding objective is the addition of firm capacity to the system, and for those in which the early months would be important for their contribution to the develop-

ment of plant units. Indeed it could well be that such new units would become, not the basis of a complete power station, but rather the last unit of a four-unit station using equipment preferably of common manufacture. By the establishment of discipline over a group of stations it must be possible to reduce the overall time for construction by an amount that is economically worth while. One of the difficulties in discussing the interim or milestone dates within a programme in terms of probability is that it always seems that the only possible deviation is 'lateness', so that the distribution of probability is completely skew. This implies that there is no incentive to the contractor to reach milestones early. If this is true it is because that is the way the contract is written. If there is a genuine belief that the 'milestones' are important, then the contract must reflect this in the only way that is recognizable by the contractor. The nature of the 'milestones' must be very clearly identified, and in the case of design work, a contract separated from that for equipment supply is often essential. It must have its own programme. It is unwise to associate Research and Development work as such with a specific project, but if test or development work is essential to prove the design, then the project management itself must include competence to assess such work. In the absence of this, to leave the matter to a separate R & D organization is to court trouble.

CONTRACT MANAGEMENT

Only when the maximum possible clarity has been achieved in the programme of work is it practicable to formulate a reasonably close budget and settle contracts.

The selection of contractors is, of course, a subject in itself involving technical, commercial, and political issues which we will not discuss here. The project management cannot be overridden in its selection of contractors without weakening their grip and chances of success. They are in an unenviable position if required for any reason to list untried contractors for any part of the works. They are not helped when established contractors go out of business or are taken over in the course of a contract, for such events can play havoc with the control of work in progress.

The terms of the contracts must be such that the contractor can get ahead with the work with a full sense of responsibility, but his work will be subject to technical audit by or on behalf of the client. The greater the extent to which this can be done by face to face discussion of the job (with appropriate record keeping), and the less by correspondence, the better. Discipline must be exercised in asking for alternative schemes to be prepared. This can cause much delay and is a hidden cause of extra costs and weakness of design work, for design teams are in general under strength.

All discussion with the contractor in relation to the contract is of a contractual nature and must be admitted as such and must be controlled by the project team, with variations (kept to the absolute minimum) promptly acknowledged and paid for. Many contractors have developed the habit of delaying such discussions, believing it to be advantageous to do so, but the project management must retain control over this.

There is probably room for a real reconsideration of contract form, objectives, and indeed language. The forms which have become hallowed by reason of legalistic phrasing do not always reflect the true objectives of the purchaser. He is a lucky purchaser who can count on getting more than he pays for, and it is perhaps in the design of contracts and specification of methods of working that the greatest progress immediately available in project management will lie. We have indicated that specific attention, specific payment, may be given for design effort if we believe this is in specific cases useful. We have said that the project team must control all design work, and the contract should be clear as to how this is to be done. Some will argue that close inspection of detail design work by the purchaser is essential. Although in some cases he has insisted on this in widely varying fields, the writer believes this to be a hazardous course, preferring concentration on the auditing of design management and rigorous final inspection of the product.

There is much discussion of the quality and content of design, and it is strongly suggested that much greater attention must be paid to methods of working in manufacture, erection, and commissioning. Manufacturers seem reluctant to send their designers and draughtsmen to sites to get the necessary experience, and so it may be necessary to send men into their drawing offices to explain these points. The recent NEDO report on large construction sites made a number of useful points about the administration of such sites, but largely overlooked the impact of design work on site labour relations. The cable races at Fawley may seem over-generously large and expensive, but they were deliberately designed for the ergonomics of cable installation rather than for minimum volume.

Though suggesting that bad design work can provoke bad industrial atmosphere on site, it cannot be held that good design work of itself gives good labour relations. The customer may, if he wishes, proclaim that labour relations are solely the preserve of the individual contractors. *Caveat emptor!* The purchaser who acts solely on this belief, doing nothing to ensure that contractors are compatible, and have compatible payment policies, should look to his contingencies and completion dates. It may be possible to succeed with a hotch-potch of incentive bonus schemes, but the purchaser who honestly comes out of the grass and admits to a real interest in site sociology, labour relations, and particularly site safety will, if he acts cautiously and sensibly, find his efforts repaid. The management skill deployed on site by contractors is still non-uniform, being in general far better for civil contractors than others, but the customer again is entitled (if his contract is properly drawn) to insist on adequate management strength. This is not always available in the con-

tractors' contracts organization, but the selection of contractors can take this into account. The site manager should be in a position to insist on an adequate build-up of management strength before any manual workers are set on. Probably, in the U.K., the customer's site organization is somewhat too large, and not given enough technical strength or authority. The time may come when a site manager who does not have an executive welding engineer/metallurgist on his staff will be thought of as unprofessional, yet at present costly errors continue to be made while the metallurgist prefers, with one or two very effective exceptions, the retreat of the R & D atmosphere.

As has been emphasized in previous papers by the author, the project management has to organize a team of teams; nowhere is this more obvious than on site, and in no sense more significant than in attention to safety matters. Though good work has been done there is room for progress. In our general standard of education and in our consequent behaviour patterns we all have some liability to accident. The discussion of this topic should be adequate, and so should the training. There is confusion in the minds of some managers, usually those most lacking in self-confidence, as to the wisdom—some would say the practicability—of including manual workers in such discussion and in site safety committees. This is to confuse negotiation and bargaining on the one hand, with organizing work and communication on the other. There are difficulties and room for improvement on trade union and management sides alike. In discussion of accidents there is, perhaps, too much concentration on personal injury. If any accident, any dropped component which could have caused injury, were treated equally, we might make more progress.

THE PLANT AS A WHOLE

So far we have spoken of design as though directed solely to bringing the plant into being. Of equal importance is the design analysis which gives an understanding of its mode of behaviour and the control of its operation.

As the benefits of fully automatic control are more generally understood, the importance of a clear understanding of plant behaviour will be more widely appreciated, and technique further developed.

There is no doubt in the writer's mind that much lack of availability must be attributed to carelessness on the part of project management who are too ready to produce a plant without developing the necessary detailed grasp of its behaviour to be able to instruct the operators. Moreover, one finds a too late specification of the control system, leading to non-availability of instruments at the time of start-up. Plants which are started up without full and completely understood instrumentation have but a poor chance of establishing good availability.

The technique of automatic control, once it is established, seems likely to advance rather rapidly, and it is to be expected that as operators become increasingly aware of its advantages it may be possible for them at least to accept that this is the most reliable way of operating a plant, recognizing that the computer and its peripheral equipment can be made just as reliable as required to make negligible contribution to plant outage, while at the same time ensuring safe shutdown when this is the appropriate action.

It is not possible, either in a book or in a short note, to give a guide to the handling of a power station project, and the writer is very conscious that it has been possible to touch on but one or two topics which, however, should assure controversy in areas in which discussion may prove useful to the art.

C84/71

AUTOMATION OF
LARGE THERMAL POWER STATIONS

E. B. JOHNSON*

On-line computer systems can, in principle, carry out many functions in power stations, ranging from simple data recording to full automation of start-up and on-load control. The current practice, as described in the paper, represents an intermediate position between these two extremes and is the result of detailed technical and economic appraisals of a wide range of possibilities. Developments in the design of conventional modulating control systems are reviewed, stressing the increases in the load range which can be achieved, the use being made of simulation in the selection of control system configurations, and the formulation generally of detailed design and performance requirements for automation schemes.

INTRODUCTION

EARLY ATTEMPTS at automation of thermal power stations in a number of countries, including the U.K., were based on the use of large on-line digital computers to control automatically the start-up and shut-down of the generating units and their auxiliaries. In most cases the computers were programmed to perform certain additional data processing and recording functions, but the primary purpose of the installations was the full automation of start-up and shut-down.

It is now generally recognized that these installations have not fulfilled the original expectations of them and that they have proved to be expensive in first cost and difficult to put to work successfully without undue interference with the normal operation of the plant. Moreover, because of the complexity of the systems, back-up manual control facilities have had to be provided, and this provision has reduced the incentive to expend the time and effort required to keep the computer control systems in working order.

These experiences have led the Central Electricity Generating Board (C.E.G.B.) to make a critical reassessment of the role of on-line computers in thermal power stations, and of the extent to which the automation of start-up and shut-down can be technically and economically justified. This paper describes the arrangements provided in C.E.G.B. stations recently commissioned, or under construction, and discusses the considerations which have led to the adoption of the current practice. For the benefit of overseas readers, the operating requirements imposed by the C.E.G.B. generation and transmission

The MS. of this paper was received at the Institution on 24th November 1970 and accepted for publication on 21st January 1971.
* Section Head, Control and Instrumentation, Central Electricity Generating Board, Walden House, 24 Cathedral Place, London, E.C.4.

system are also described in so far as they affect the design of control and automation systems.

The success of power station automation schemes depends ultimately on the reliability and performance of a very large number of items of equipment for the detection and regulation of plant operating stations. Many of these devices are necessarily located on or around the plant and have to operate under arduous environmental conditions. Some of the arrangements within the C.E.G.B. for evaluating the suitability for power station application of commercially available instrumentation and control equipment are described.

SYSTEM OPERATION REQUIREMENTS

The increasing amount of nuclear generation at present only suitable for base load operation which is being installed in this country makes it necessary for all conventional generating plant to be designed to be suitable for two-shift operation, i.e. for daily starting up and shutting down. Some of the 500-MW units now being installed are expected to operate on a two-shift regime at certain times of the year very early in their lives.

A typical curve of the daily load on the C.E.G.B. system in summer shows the demand starting to fall sharply at about 23.00 h and rising again rapidly in the morning at 06.00 h. Duration of shut-down periods of plant that is two-shifted is, therefore, typically between 6 and 8 h. Longer periods of shut-down of the order of 36 h occur at week-ends.

The morning increase of load on a winter day is about 13 000 MW over a period of 2 h. It is therefore important that individual generating units should be able to pick up load rapidly. In practice, the rate of pick-up is determined by the characteristics of the generating plant and the ability of the boilers to supply steam at the temperature required to match the thermal condition of the turbines at start-up. C.E.G.B. specifications require plant to be capable of increasing output to maximum rating in a period of 20 min from synchronizing after an overnight shut-down.

Generating plant is also required to be capable of providing spinning capacity to meet sudden demands for increased output resulting from system faults. The most severe system fault against which provision is currently made is the simultaneous loss of two 660-MW units, and it has been calculated that at the time of summer light load, when the system inertia is at a minimum, such a loss would cause the system frequency to fall initially at the rate of about 0·75 per cent/s. In the operation of the system at present, spinning spare capacity to meet such demands is normally carried or plant loaded to 75 per cent continuous maximum rating (c.m.r.). Sufficient additional output to sustain the system must be available within a few seconds in order to prevent the frequency falling below 48·5 Hz, at which point low-frequency load-tripping relays operate.

PLANT OPERATING CONCEPTS

The loading of generating units is carried out by the adjustment of turbine governor set points. The change in turbine steam demand is seen by the boiler as a change in steam pressure, and this causes the boiler automatic control system to adjust the heat input accordingly. Consideration has been given in the past to the use of the alternative system, in which load demands are applied to the boiler heat input and the turbine governing valves are controlled to take the steam generated by the boiler. This system tends to improve plant operating conditions by maintaining a better balance between boiler output and turbine demand. The system has not been adopted because it reduces the ability of the unit to respond quickly to sudden demands for increased output, as described above.

Plant parameters, such as pressure, temperature, and water level, are kept as constant as possible during normal load changing. However, almost all the C.E.G.B.'s large generating units have drum-type boilers with natural or assisted circulation. For reasons of economy, the specifications for these require the main and reheat steam temperature to be maintained at the design value only over the load range from 70 to 105 per cent c.m.r. (Note: 105 per cent c.m.r. is obtained by by-passing the high-pressure feed water heaters and is intended only for short-term use at the time of system peak load demand.) Below 70 per cent load the steam temperature falls off according to the natural characteristic of the boiler.

At start-up after an overnight shut-down the temperature of the steam from the boiler tends to be too low to match the turbine metal temperature. To overcome this difficulty, the start-up technique usually adopted is to commence firing with the boiler drum pressure reduced to a value well below the normal operating level and to raise the drum pressure progressively during loading. The overfiring necessary to raise drum pressure has the effect of increasing the main and reheat steam temperatures above the steady-state values corresponding to the prevailing steam flow, thereby producing a better match between steam and metal temperatures. For success, this technique requires a fairly rapid and uninterrupted start. If a hold-up occurs for any reason, firing has to be reduced to prevent boiler pressure rising excessively, and the temperature tends to fall off as a result.

EXTENT OF AUTOMATION OF START-UP AND SHUT-DOWN

Current C.E.G.B. practice is to provide a central control room for each station, from which all operations required for routine start-up, shut-down, and on-load control of the generating units and associated unit auxiliary plant can be carried out. The installation is designed with the objective of permitting each unit to be started up after an overnight shut-down by a single operator in a period of 1 h from light-up of the boiler to full load.

To establish the extent of control and instrumentation, and the degree of automation required to enable the above objective to be achieved, a critical path study of the operations required for a hot start of a 500-MW coal-fired unit has been carried out. Fig. 84.1 shows a small section of the critical path diagram resulting from the study.

The critical path diagram is divided into five phases,

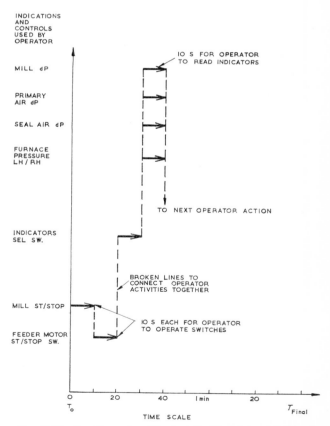

Fig. 84.1. Section of critical path diagram of boiler/turbine hot start procedure

namely: (1) initial start-up; (2) pressure and temperature raising; (3) turbine run-up; (4) synchronizing; and (5) unit loading.

Examination of these phases indicates the work load on the operator and shows that the operator's busy periods are during the initial start-up, synchronizing, and the beginning of the unit loading phases. The pressure- and temperature-raising phases comprise a slack period by comparison.

The study has shown that the minimum degree of automation required to enable the target period of a 1-h start by a single operator to be achieved is:

(a) Automatic sequence control of unit auxiliaries, a separate sequence being provided for each functional group of plant items and each sequence being initiated manually by the operator at the appropriate stage of the start-up procedure.

(b) Automatic turbine run-up to speed.

(c) Automatic turbine synchronizing.

(d) Automatic turbine loading.

(e) Automatic ramp loading of boiler heat input, the slope of the ramp being selected by the operator to match the rate of turbine loading.

(f) Extension of the range of operation of the boiler modulating controls to cover lower loads.

In addition to reducing the work load on the operator, and so enabling the target start-up time to be achieved, this automation has the following advantages:

(1) Elimination of the individual control switches required for starting up or operating each plant item enables the size of control desks to be reduced and their ergonomic design improved.

(2) Turbine supervisory measurements are automatically monitored by the automatic run-up and loading system, and the risk of damage reduced.

(3) The use of uniform predetermined loading rates improves the overall control of the boiler–turbine generating unit.

(4) It provides a facility for carrying out normal load changes, either manually or automatically.

DESIGN OF SYSTEMS FOR AUTOMATIC START-UP AND SHUT-DOWN

Automatic sequence control

Separate start-up and shut-down sequences are provided for each of the major auxiliary plant groups. A typical installation would include sequences for the following: (a) each forced-draught fan, (b) each induced-draught fan, (c) each air heater, (d) each coal pulverizing mill, (e) gland steam supply and vacuum raising, (f) condensate system, (g) starting and stand-by boiler feed pumps, and (h) each circulating water pump.

A sequence comprises a number of sequence steps, each of which is concerned with a plant item. A sequence step consists of a command signal to a plant item, followed by a completion check signal which allows the next step to be

initiated. Once initiated, each sequence should proceed automatically to completion within a predetermined time interval, which is monitored by a timing unit. If a sequence fails to complete within the specified time, the timing unit initiates an alarm and inhibits further progress of the sequence. Where a modulating control loop is associated with a plant group, completion of the start-up sequence leaves the plant operating on automatic control, without the necessity for manual intervention. Interconnection between the sequence and modulating control systems is necessary for this purpose. Fig. 84.2 shows a typical sequence flow sheet for the start-up of an induced-draught fan.

Automatic run-up and loading

The equipment is designed to accelerate the turbine from turning gear speed to a speed within the range of the synchronizing equipment, by operating the emergency stop valves. It loads or unloads the machine to a value selected by the operator, by operating the governing valves via the turbine governing system.

The equipment is divided into three sections.

Pre-start checks

A series of checks on the state of the plant is carried out to determine whether it is safe for run-up to take place. Only if these checks are satisfied can the run-up stage be entered.

With one particular design, the equipment also automatically selects the optimum run-up rate, the optimum value of block load, and the optimum loading rate and displays these to the operator. These optimum rates are calculated on the basis of turbine casing metal temperature immediately prior to the admission of steam. Selector switches are provided, however, to enable the operator to select different rates from the optimum, should he think fit.

Run-up

Run-up rates corresponding to run-up times varying between 5 and 60 min are available to suit varying plant conditions. In the majority of systems being installed, run-up is effected by means of a closed-loop servo system controlling the opening of the turbine stop valves in relation to the turbine speed. The equipment generates a ramp function in a static digital counter which is converted to an analogue output voltage representing the desired value of turbine speed. An error signal corresponding to the difference between the desired and actual speeds is used to control the opening of the turbine stop valves. Protection against malfunctioning is given by independent circuits which detect excessive rates of change of ramp output and excessive error between ramp output and turbine speed.

Turbine supervisory measurements such as vibration, eccentricity, differential expansion, and thermal stress are monitored by the run-up equipment. The programme is

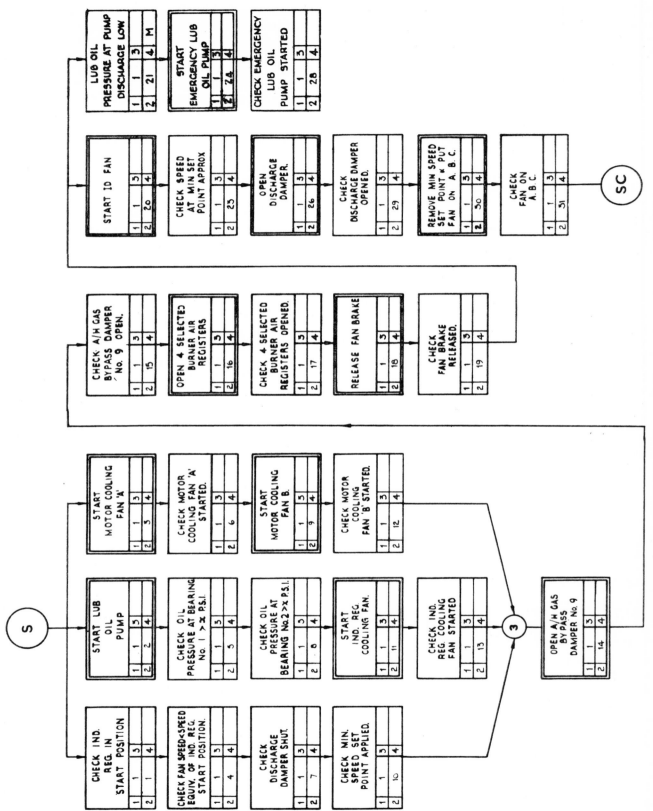

Fig. 84.2. Typical sequence flow sheet for start-up of an induced-draught fan

automatically modified or inhibited in the event of pre-determined values being exceeded.

Turbine loading and unloading

Block loading following synchronization is carried out by the equipment, and block loads equivalent to 5, 10, and 20 per cent are available for selection. Subsequent loading to the target load, for cold or warm starts, is carried out in three stages—up to 20 per cent c.m.r., 20–50 per cent c.m.r., and 50–100 per cent c.m.r. A choice of loading rates for each stage is available, varying between a minimum of 0·1 per cent c.m.r./min in stage 1 and a maximum of 10 per cent c.m.r./min in stage 3. For hot starts it is planned to use a single loading rate, from block load to target load, of 5 per cent/min. The operator is provided with a target load control which enables him to set target load between zero and full load in 10-MW steps.

The load control loop takes the same form as the speed control loop described above, and protection against malfunctioning is similarly provided by independent circuits detecting excessive error signals. Turbine supervisory measurements are again monitored by the equipment, and the loading programme is halted in the event of predetermined values being exceeded.

Boiler loading

During the start-up period the fuel input to the boiler must meet the demand imposed by the turbine loading, and in addition build up boiler pressure from the level to which it has fallen during the shut-down period. On a coal-fired plant the output of each pulverizing mill is regulated by means of a primary air flow control loop which forms part of a total fuel control circuit. Each mill is started by an automatic sequence control scheme after manual initiation by the operator at the appropriate time. The start-up sequence for each mill concludes with the automatic transfer of the primary air regulator to control by the total fuel controller. The facilities provided on the central control desk include provision for selecting three alternative modes of operation of the total fuel controller, namely auto, manual, or ramp. The last of these provides for a ramp input to the desired value of the total fuel controller at a rate which can be manually adjusted.

Up to, and including, the stage at which a block load is applied to the turbine, total fuel control is carried out manually with one or two mills in operation, at well below their maximum output. Immediately following the application of block load, the total fuel controller is switched to the ramp position. From this point onwards the desired value of the total fuel controller is increased at a selected rate which can be adjusted or overridden at any time by the operator. As the mills, which are on load, approach their maximum output, further mills are brought on load by the operator initiating the appropriate automatic sequences.

The ramp input to the desired value of the total fuel controller is allowed to continue until the normal desired value of the steam pressure is attained, whereupon the control is switched from the ramp to the auto position. Thereafter, control of mill output is accomplished by the master steam pressure controller via the total fuel controller.

ON-LOAD UNIT CONTROL

The boiler control scheme is designed to enable the boiler to meet the load demands imposed upon it by the turbine, whilst maintaining adequate quality of control of certain boiler variables. These load demands result from the response of the turbine speed governor to system frequency changes and from load changes brought about by the action of the operator, or the automatic loading equipment, on the set point of the governor.

On later units a continuously acting load control loop will be provided. This load control loop is relatively slow in operation, employing integral action only and operating on the set point of the turbine speed governor. The immediate response of the turbine to changes in system frequency is thus determined by the speed governor, with corrective action being applied more slowly by the load control loop.

The load loop measures generated power and compares this with a desired value derived from the datum value of load required at 50 Hz, plus or minus an amount proportional to the prevailing frequency deviation from 50 Hz. The load controller's integral action is applied to any error resulting from this comparison. The extent to which a deviation in frequency changes the desired value input to the load controller depends upon the 'load-loop droop' setting. A choice of values of this is provided to suit changing system operation requirements. In addition, a 'non-regulating' mode of operation is available in which no change in desired value of load occurs, unless the frequency deviates outside a predetermined range.

The control techniques applied in boiler-modulating control schemes depend considerably upon the particular plant design features involved. It is beyond the scope of this paper to deal with them in any detail. However, it is worth emphasizing the influence which current trends in automation and closer definition of operating requirements are having on their basic design. Automatic start-up, wider load ranges, and the provision of spinning spare capacity all considerably increase the requirements imposed on the performance of the boiler-modulating control system. These not only demand a more rigorous approach to the design of the system, and the choice of the equipment required to implement it, but also justify greater value being given to the 'controllability factor' in the design of the boiler plant and the selection of its auxiliaries.

The need for a more rigorous approach to control system design has resulted in much effort being applied in recent years to the use of simulation techniques, in the design stage of new projects, to examine the dynamic response of the plant and to assess the performance of proposed control systems. Use is being made of analogue,

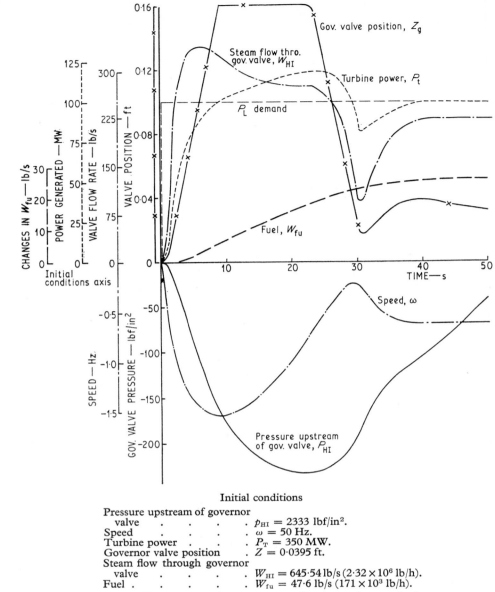

Fig. 84.3. Initial transient response of main plant parameters to a step demand in load (P_L)

Initial conditions

Pressure upstream of governor valve	.	.	.	$p_{HI} = 2333$ lbf/in².
Speed	.	.	.	$\omega = 50$ Hz.
Turbine power	.	.	.	$P_T = 350$ MW.
Governor valve position	.	$Z = 0.0395$ ft.		
Steam flow through governor valve	.	.	.	$W_{HI} = 645.54$ lb/s (2.32×10^6 lb/h).
Fuel	.	.	.	$W_{fu} = 47.6$ lb/s (171×10^3 lb/h).

N.B.—The responses shown represent the change in magnitude
of the parameters from their 70 per cent c.m.r. value.

digital, and hybrid computing facilities for these studies. In addition, dynamic response tests are being carried out on newly commissioned plant to support the theoretical work. Fig. 84.3 shows some of the results of a study of the ability of a 500-MW oil feed unit to respond to sudden load demands caused by falling frequency following a loss of generating capacity on the system.

ON-LINE DATA PROCESSING AND COMPUTER CONTROL

A functional and economic analysis of the application of

data processing and computer control to thermal stations has been made. The analysis was based on a centralized system for a four-unit station, the following functions being considered.

Routine data logging

In existing large stations approximately 200 plant parameters are recorded by hand for station management purposes. A data logger or computer can record these readings and provide immediate warning of their going outside pre-set limits. Readings can be taken over a much shorter time than is possible by manual logging, are free from operator 'bias', and continue during periods of difficult operation when human operators are forced to discontinue logging.

With the intended manning of new stations the employment of additional staff to cover manual logging would be necessary. There is considered to be an economic justification for installing a data logger for this purpose, unless the function is carried out by a computer performing other tasks as well.

Alarm print-out

This provides a typed list of alarms showing times of operating and resetting. It has the advantage of giving a permanent record of alarms, together with their order of occurrence, thus providing a degree of analysis of the cause of incidents. The relatively slow print-out, however, requires the retention of existing alarm facias.

Alarm display

A further possibility is to arrange for alarms to be written out on cathode-ray tube displays in place of the conventional alarm annunciators. Some conventional alarms have to be retained to avoid the need for the plant to be shut down in the event of a computer failure.

Alarm analysis

With cathode-ray tube display an analysis of alarm situations can be included. This can vary from the simple suppression of 'standing' alarms, or alarms inappropriate to the plant condition, to an analysis of the 'prime cause' of the alarm situation.

A study of conventional alarm schemes on existing thermal stations has not produced any evidence of plant damage being caused by the inability of operators to interpret conventional displays. The tentative conclusion is that there is little advantage in going beyond suppression of standing alarms and display in order of occurrence.

Trend monitoring

Trends of measured variables can be displayed in graphical form on a cathode-ray tube. An advantage is that an expanded time scale equivalent to a fast chart speed can be used. Some financial saving can be obtained by the elimination of conventional recorders. However, the extent of this is limited by the need to provide 'back-up' for computer faults.

Current data display

Conventional instrumentation can be replaced by analogue, digital, or graphical displays on cathode-ray tubes. There are advantages and disadvantages as compared with conventional instruments, and there is insufficient experience to determine where the balance lies. Cost savings from elimination of conventional instruments are small.

Post-incident recording

This is considered to be a valuable facility for the investigation of the causes of incidents leading to loss of output. Readings of selected parameters are stored every minute, for the preceding half-hour, or at shorter intervals where required. In the event of a fault, these are printed out for subsequent analysis. A distinct advantage of the system over conventional chart recorders is the accuracy of the time scale, which permits accurate lining up of associated parameters. Computer records are, however, normally available only in tabular form, which is not very convenient for analysis purposes. The provision of automatic graphical plotting equipment may cost more than the saving from the elimination of conventional recorders.

Plant performance analysis

The facility, provided by a computer, for rapid calculation of thermal efficiency and individual plant item performance should provide an early warning of loss of performance. In addition, it should enable plant operation and maintenance to be adjusted to improve thermal efficiency. In the absence of operational experience, however, the assessment of the financial return for such schemes is extremely difficult. Instrumentation accuracy appears likely to be the major problem.

Statistical calculations

Certain routine statistical calculations are required to be made from recorded data for management purposes. A small financial saving can be achieved by the elimination of the clerical effort normally involved.

Turbine run-up and loading

It is the policy of the C.E.G.B. to provide automatic run-up and loading of turbo-generators at all future thermal stations. Purpose-built equipment has been provided for this purpose at a number of recently commissioned stations, but in later stations the function will be transferred to a station computer. No stand-by computing equipment is required, as back-up manual facilities are available, and a significant financial saving can therefore be achieved.

Continuous control of turbine output with adjustable frequency–load characteristics

Some savings in system operation can be achieved by providing a facility to enable the turbine load–frequency characteristics to be adjusted to suit changing operational circumstances. This will enable short-term unpredicted consumer demand fluctuation to be distributed among the units operating on the system in a more economical manner than at present. This facility, which can best be provided by closed-loop control of turbine output operating on the set point of the turbine speed governor, is being included in computer schemes for new stations.

Auxiliary plant sequence control

At present, fixed-wire systems for sequential start-up and shut-down of plant auxiliaries are being provided. Since an entirely independent sequence is being provided for each auxiliary group, and no single fault can affect more than one sequence, no back-up provision, other than that of local control of switchgear, is being made to cover sequence failure. If sequence control is carried out by a central computer, then, to provide a comparable degree of reliability, redundancy must be included in the computer system. The minimum degree of redundancy considered to be acceptable is two central processors with back-up stores, etc., and input–output units segregated into about 10 modules per unit.

Unless this degree of redundancy is being provided for other reasons, the cost of the computing installation is increased considerably. Thus, there appears to be no overall financial advantage in transferring the sequence control from fixed logic to the computer.

Modulating control

Continuous on-load control of plant parameters, which is at present carried out by conventional process controllers, could be performed by direct digital control from the computer. However, on a modern large unit, simultaneous loss of all modulating control loops would probably require the plant to be shut down, and the computer system for such a function would need to have an extremely high reliability. It is considered unlikely that such a system could be economically justified at the present time.

The cost of computing systems required to carry out various combinations of the functions described above has been assessed. As a result, the following conclusions, relating to a station comprising several large generating units, have been reached.

(1) There is a fairly clear economic justification for a central computing system to carry out routine data logging, post-incident recording, alarm print-out, statistical calculations, turbine run-up and loading, and the control of turbine output with adjustable load–frequency characteristics.

(2) The addition of comprehensive cathode-ray tube facilities can increase the cost of the installation to an extent which it is difficult to justify. Recently, however, much cheaper equipment for providing tabular format displays has become available, and this appears to be economically attractive for the presentation of alarm messages.

(3) The potential savings from plant performance analysis are considerable, but experience is needed to determine whether these can be realized in practice.

(4) The addition of sequential control of auxiliaries, or modulating control of plant parameters, to a computer system carrying out any or all of the functions indicated above would require the provision of redundancy in the computer system, to enable an accept-

able reliability to be achieved. It does not appear that such a system could be justified economically at the present time.

CONTROL ROOM DESIGN AND OPERATOR FACILITIES

Current practice is to provide a single control room for the station from which all routine starting up, on-load control, and shutting down of the generating units can be carried out. The control room includes the following equipment.

(a) A control desk and associated instrument panels for each generating unit.
(b) Control panels for high-voltage switching.
(c) Control panel for auxiliary electrical supply system.
(d) Cooling water system control panel.
(e) General services control panel.
(f) Fire service panel.
(g) Supervisory engineer's desk.

For unit control the preferred arrangement is a control desk backed by a separate vertical panel. The vertical panel contains the following items.

(a) Equipment for pre-start operations and checks prior to boiler light-up.
(b) Equipment for remote manual control of selected plant items under fault conditions.
(c) Alarm display.
(d) Master instruments and additional instruments related to the control desk.
(e) Recorders and other instruments not required for minute-to-minute operation.

The control desk contains the equipment required to enable a single operator to carry out the following procedures.

(a) Hot start-up of the generating unit from light-up to full-load operation.
(b) Normal on-load operation, including control of generated power level.
(c) Normal shut-down.
(d) Emergency action to safeguard the plant in the event of a fault.

For the earlier 500-MW units, the length of the unit control desk is of the order of 30–40 ft, making it impracticable for a single operator to control and supervise the plant effectively. At the later stations a very much more compact design of control desk, as illustrated in Fig. 84.4, has been made possible by the introduction of sequence control for start-up of unit auxiliaries, multiplexing of auto–manual control stations, and the use of smaller equipment and more compact layout.

The use of sequence control not only reduces the work load on the operator during start-up but also reduces the number of control switches required to be mounted on the unit desk. Facilities for switching the individual plant items, covered by the sequences, are not normally

Fig. 84.4. Cottam power station control room

provided in the control room, but would be mounted on the back panel if required for any particular purpose.

On the largest coal-fired boilers now being installed by the C.E.G.B., up to 10 fuel pulverizing mills are provided. Each mill has four modulating control loops associated with it, giving a total of 40 loops for the complete milling plant. To mount the auto–manual control stations for this number of loops on the control desk would add considerably to the length of the desk. To avoid this, only two sets of control stations are provided with facilities for each of these to be switched to any mill. The use of the raise-lower type of control system, in which the controller output is zero when the plant is in a steady-state condition, facilitates this multiplexing.

For the 660-MW units at Drax a modular system of instrumentation for the unit control desk has been developed. In this system, the control desk is virtually a framework of racks designed to receive basic instrument modules. The modules are all of the same length and their widths are multiples of a fixed modular dimension. The major part of the desk layout is covered by three types of modules. These are an indicating instrument module, a push-button control module, and a discrepancy-switch control module.

For future stations it is hoped that suitable indicators of the vertical edgewise type will be available for the instrument module. However, at Drax, circular indicators are being used. The push-button control module can accommodate up to eight push buttons. Together with an appropriate instrument module, it replaces the proprietary auto–manual control station used on previous installations. The discrepancy-switch module is used for the initiation of auxiliary plant sequences and other switching operations. The modules are connected from the desk to the cable-marshalling rack by preformed factory-assembled cables, terminated by standard plugs and sockets.

Fig. 84.5 shows a pictorial representation of the Drax control desk, which uses modular instrumentation and

Fig. 84.5. Drax power station control console unit

control. This desk will enable a single operator to start up, supervise, and control the plant, using 20 start-up and shut-down sequences; 80 modulating control loops; automatic turbine run-up, synchronizing, and loading; automatic control of unit output with compensation for system frequency variations; and data processing and presentation, utilizing cathode-ray tube display.

The advantages of the modular system follow.

(1) Only the outline dimensions of the desk frame need to be settled at an early date. No complex panel cut-outs are required and the instrument array can be modified at any time, even after commissioning, without incurring additional costs or delays and without detriment to the ergonomic design of the installation.

(2) A consistent practice can be followed in the arrangement of equipment on the desk, e.g. control initiation is always to the right of the indication of the controlled variable. This is not possible for cascade control loops if conventional auto–manual control stations are used.

(3) The arrangement of the controls on the desk can readily be related to the plant process flow. To illustrate this, Fig. 84.6 shows the process flow sheet for draught plant control, with the corresponding control module shown beneath it.

(4) The modular system enables special-purpose controls to be presented to the operator in the same form as conventional controls. For example, the desk shown in

Fig. 84.5 contains a direct digital control scheme for turbine loading. The raise–lower auto and manual controls take exactly the same form as those for the more conventional draught plant modulating control loops.

(5) Instrumentation and controls can be accommodated on the desk at a higher density, thus enabling a reduction in desk size to be achieved without the use of miniature instruments, which are ergonomically unsatisfactory. Instrument scales of approximately 10 cm length are used, thus permitting 1 per cent reading accuracy by an operator in the centre of the desk area at approximately 120 cm distance from the instruments.

EQUIPMENT PERFORMANCE AND RELIABILITY
Evaluation of commercial equipment

The C.E.G.B. operate a formal approvals scheme for the evaluation of commercial instrumentation and control equipment. At the present time, the scheme covers most of the more commonly used measuring devices and transmitters, and it is gradually being extended. For each type of equipment covered by the scheme a detailed specification is prepared, defining the requirements for performance, design, construction, and testing. Before approval, the specified range of tests has to be carried out either by the C.E.G.B., an independent testing authority, or in the

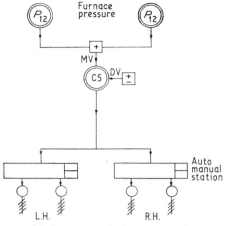

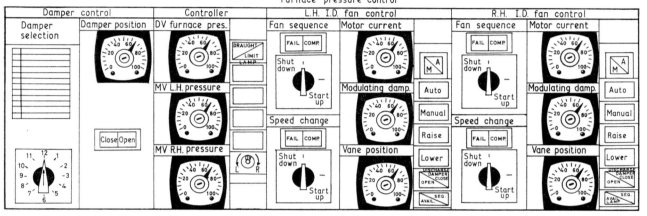

Fig. 84.6. Modular arrangement of furnace pressure control

manufacturer's own laboratories. In the latter case, the manufacturer's testing facilities have to be approved, and the tests are witnessed by a C.E.G.B. representative. The list of tests specified covers all aspects of performance which are of interest, according to the type of equipment involved.

For example, for pressure transmitters the following characteristics are tested, to determine whether they meet the specification requirements: dead zone; output ripple; effect of output terminal earthing; interference rejection; pressure leakage; effect of supply voltage and frequency variation; resistance to pressure overload; effect of dry heat, damp heat, and low temperature; effect of variations in mounting position; ability to withstand vibration; start-up drift and long-term stability; and frequency response.

Instrument inaccuracies

Since accuracy requirements vary according to the particular application, a number of different categories of accuracy have been included in the individual specifications referred to above. For example, the specification for pressure transducers includes the following categories.

Category 1	±2 per cent of the span.
Category 2	±1 per cent of the span.
Category 3	±½ per cent of the span.
Category 4	Special, to be agreed.

For temperature transmitters, on the other hand, only two categories are included, corresponding to 0·1 and 0·5 per cent of span maximum permissible error.

Ranges and scales

Ranges of instruments are usually such that the normal working point corresponds to about 75 per cent of full-scale deflection. On many installations, certain instruments measuring the main operating parameters of the plant are designated as 'master' indicators, and they are grouped together at the centre of the instrument panel. These instruments are of the circular type and are so arranged that, as far as possible, the normal working point is at the 12 o'clock position on the scale. This enables departures from normal working conditions to be readily detected by the operator.

Current practice in the design of instrument scales

follows the recommendations of British Standard 3693, Part I, *The Design of Scales and Indexes,* with certain minor additions. This Standard gives recommendations for choice of scale lengths for readability at different distances, preferrred scale ranges, division of scales and numbering of scale marks, dimensions of scale marks, and design of figuring and lettering.

Actuators

Hydraulic, pneumatic, and electric actuators are all considered to be technically acceptable, provided that they are used in applications suited to their individual characteristics. General requirements for actuators are that they should be provided with means for local hand operation, be fitted with devices to lock them in their existing position on loss of the operating medium, be protected against the ingress of dust, and suitable for operating in an ambient temperature of 65°C. For feedback of actuator position, the use of an electric transmitter of the variable transformer type is specified. The speed of response of actuators is selected to suit each particular application. Hydraulic actuators have been found to be the most versatile. They respond smoothly and can be designed to deliver the required operating torque without mechanical gearing. Non-inflammable fluids of the phosphate-ester type are being used on a number of installations. Some difficulties were experienced initially, owing to a lack of appreciation of the precautions that need to be observed, but it is believed that the problems have now been overcome.

Pneumatic actuators have been found to be very reliable, except where installed in locations having an excessively high ambient temperature. They are, however, considered to have insufficient positioning accuracy for applications involving high-inertia loads.

We have less experience with electric actuators for modulating control than with the other types; but where troubles have been experienced, these appear to have been due either to underrating of the motors or to the use of contactor-type starters. Our current requirement is for solid-state control gear to be provided.

Reliability and system redundancy

The principle which is followed, as far as possible, in the design of measuring and control systems is that a single fault should not result in the plant having to be shut down. On this basis, the loss of a control loop, or several control loops, is acceptable, provided that adequate facilities remain for the operator to take over manual control. The simultaneous loss of all control loops would not, however, be acceptable, because it is doubtful whether manual control would be practicable under such circumstances, owing to the large number of control loops involved.

To meet the above design criterion, duplicated measurement of all essential plant variables is required. This ensures that if a control loop is lost due to failure of the measuring transducer associated with it, a separate indication of measured value will be available for manual

control purposes. In practice, this does not mean that every measuring transducer has to be duplicated, since in many cases associated measurements are available, which are adequate for manual control. For example, where superheater outlets divide into two it is normal to measure the temperature in each line and use the average value for automatic control. Therefore, if one transducer fails, an indication from the other would be sufficient for manual control purposes, whilst repairs were being effected.

Redundancy is provided in compressed-air and electrical-supply systems for instruments and controls.

Compressed air is now usually required only for supplying pneumatic actuators. However, in modern stations, C.E.G.B. practice is to provide an independent compressed-air system for each generating unit, so that a failure of one system will not affect more than one unit. In addition, each system includes stand-by compressor plant and storage receivers, capable of holding about 10 min supply of air.

Practice in the arrangements of electrical supplies to instruments and controls varies somewhat between different stations; but in all cases, nowadays, provision is made for a special 110-V a.c. supply system, provided by motor-generator sets or static invertors. These are connected to a 110-V d.c. board supplied through rectifiers from the 415-V, three-phase station services board. Batteries floating on the d.c. board maintain supplies in the event of the 415-V supply being lost. The primary purpose of this system is to ensure that electrical supplies to instruments and controls are not dependent upon external electrical conditions, and that, in adddition, the occurrence of transients due to station switching operations, which might be harmful to solid-state electronic equipments, is reduced.

COST OF INSTALLATION

The cost of control and instrumentation varies considerably, according to the type of generating plant installed and the degree of automation provided. The cost of cabling associated with control and instrumentation is high and has been found to vary a good deal between stations, depending, apparently, upon station design and location of control equipment.

For a station with three 660-MW coal-fired units designed to current automation concepts, i.e. sequence

Table 84.1. Equipment cost breakdown

Equipment	Percentage
Boiler plant	24
Turbo-generator, feed-heating and condensing plant, and boiler feed pumps . . .	12
Miscelleneous and common equipment . .	16
Sequence control, interlocking, and intertripping	10
Computer	14
Cabling	24
	100

control of auxiliaries and a station computer for data handling and automatic turbine run-up and loading, the total cost of control and instrumentation is about $3\frac{1}{2}$ per cent of the total station cost. The cost of the equipment can be broken down into areas approximately as detailed in Table 84.1.

SUMMARY OF EXPERIENCE TO DATE WITH AUTOMATION

The experience with automation, other than the conventional control and instrumentation systems, falls mainly into the following three areas.

    (*a*) Automatic sequence control.
    (*b*) Automatic run-up and loading of turbo-generators.
    (*c*) Data processing and computer control.

These areas will now be discussed separately.

Automatic sequence control

Comprehensive automatic-sequence control systems are in operation at three stations: Cottam, Rugeley 'B', and Pembroke. The first two of these are coal-fired and the third is oil-fired.

Apart from minor initial teething troubles, the experience with the sequence control equipment itself has been good. However, there have been problems with plant-mounted initiating devices, such as limit switches. It is evident that much more attention will need to be paid in the future to the selection and method of application of these devices. Despite these problems, automatic-sequence control is now accepted by operating staffs as a viable form of control.

Automatic run-up and loading

Automatic turbine run-up equipment is in operation at both Rugeley 'B' and Cottam. The equipment on No. 1 at Rugeley 'B' was the first of its type, and some minor teething troubles were experienced initially. On No. 2 unit at Rugeley 'B' the equipment was available from an early date and was, in fact, used on the second run-up of the machine. This indicates the confidence of the turbo-generator contractor and the Board's operators in the equipment. Similarly, the auto run-up equipment at Pembroke, which is of the same design, was used at a very early stage in the commissioning of the plant. Fig. 84.7 shows a recording of turbine speed and rotor eccentricity measurements taken during an automatic run-up at Rugeley 'B' on 20th July 1970.

Automatic loading equipment is also provided at Rugeley 'B' and Cottam, but plant problems have prevented it being commissioned to date.

Data processing and computer control

Cottam and Rugeley 'B' have fairly simple data logging installations and these are working satisfactorily.

At the Fawley station, each of the four 500-MW oil-

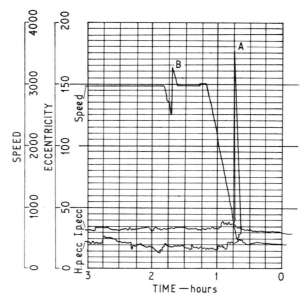

Rugeley 'B' power station, unit 6, turbine auto start, 20th July 1970.

Note

A Change in range of speed-measuring instrumentation from 0–400 rev/min to 0–4000 rev/min.
B Turbine overspeed trip test.

Fig. 84.7

fired units is provided with a computer for complete automation of start-up and shut-down, for fault detection and remedial action, and for performance monitoring, logging, and other data processing functions. At the present time, three of these computers are in use for data logging purposes only, pending completion of the writing and testing of the programmes for the remaining functions and of commissioning of the required inputs to the computers.

The Fawley computer installation is a very ambitious one, committed primarily with the object of enabling experience of computer control to be obtained, and it can be expected to take some time to commission fully. At Pembroke, the installation is much more nearly in line with the Board's current concept for the application of computer systems in thermal power stations, and consists of a central computer serving all four units in the station. The functions to be carried out by this computer are data logging; data presentation on cathode-ray tubes; alarm presentation, analysis, and recording; performance monitoring and analysis; and post-incident recording.

At the time of writing, the first unit at Pembroke is in the early stages of commissioning, and the computer system is not yet operational.

Our experience to date indicates that the provision of the necessary effort for the development, proving, and commissioning of the programmes for on-line computer systems presents a major problem. It is, therefore, considered to be highly desirable that these programmes should be kept as simple as possible and that the temptation to incorporate very complex functions should be resisted.

TRENDS AND FUTURE POSSIBILITIES

The future development of the automation of thermal power stations in this country is likely to be conditioned by two factors. These are:

(a) The probability that an increasing proportion of new stations in the future will be of the nuclear type.

(b) The consequent reduction in design load factor of the thermal stations that are built.

These factors lead to the following tentative conclusions.

(a) It is unlikely that there will be any large-scale development projects. In the past, some installations have been authorized in order to gain operational experience with advanced automation schemes, even though it was not possible to establish a clear economic justification for them. Economic criteria are likely to be more strictly applied in the future.

(b) There will be less emphasis on the value of improving thermal efficiency and even more than there is at present on the need to increase operational flexibility, i.e. to design for frequent starting up and shutting down, and the ability to change load rapidly.

The C.E.G.B.'s current assessment of the economics of thermal power station automation indicates that for a station of up to four operating units there is justification for a single on-line digital computer to carry out data logging, post-incident recording, alarm print-out, automatic turbine run-up and loading, and continuous control of turbine output with variable frequency–load characteristics. Other features which are currently being provided are alarm and data display on cathode-ray tubes and plant-performance monitoring and calculation.

It is generally believed that the cost of digital computers will continue to fall with time. The natural expectation is, therefore, that the trend will be for some of the functions at present carried out by conventional equipment to be transferred to the computer. However, with the functions carried out by the computer limited to those outlined above, it is possible, without undue expense, to design the control installation in such a way that operation of the generating plant is still possible, even if the computer is out of service for maintenance or repair. The extension to further functions, such as sequence control for start-up of auxiliaries and on-line modulating control, would necessitate either the retention of existing conventional equipment or the design of a computer installation with sufficient redundancy to make the risk of outage sufficiently low to enable possible non-availability of the plant, in the event of a computer fault, to be accepted. In either case, additional costs are involved which seem likely to make such schemes economically unattractive for some time to come.

The extent of automation being provided in the latest C.E.G.B. stations is considerable. The developments embodied in these systems need to be evaluated operationally and consolidated before further advances are made, particularly if these show only a marginal economic advantage. Experience so far suggests that the major problem will be equipment unreliability. A good deal of work is currently being carried out to improve standards in this respect, and it is expected that continuing effort will be required.

OPERATING EXPERIENCE WITH A COMBINED-CYCLE GAS TURBINE AND STEAM TURBINE PLANT AT THE HOHE WAND POWER STATION OF THE NIEDERÖSTERREICHISCHE ELEKTRIZITÄTSWERKE AKTIENGESELLSCHAFT

H. CZERMAK*   K. GOEBEL*

This paper describes operating experience with the first high-merit steam plant in Europe to use a Topping gas turbine. The reasons for selecting a combined cycle plant and the thermal cycle employed are explained. The fuel-firing system, in which the gas turbine exhaust is used as the combustion air for any of three different boiler fuels, is noteworthy; a detailed description is given. Outages and defects during the past six-year life of the station are covered and cost figures given.

INTRODUCTION

THE HOHE WAND power station in Austria was built for the state-owned NEWAG and NIOGAS companies between 1962 and 1964 and commenced base-load generation in January 1965. The principal data of the gas turbine and steam plant, which normally operate together in a combined cycle, are given in Table 89.1. These two companies are responsible for supplying the state of Lower Austria with electric power and natural gas respectively.

The erection of a thermal power station had become necessary owing to load growth, and when official operation commenced it supplied about 50 per cent of the base load in the NEWAG area. During the six years' operation until the close of 1970 the gas turbine had completed 43 156 h running and the steam turbine 42 846 h.

THE THERMAL CYCLE

Manufacturers were requested to submit proposals for a plant that would satisfy a number of unusual requirements, while at the same time exhibiting high reliability through the use of conventional, yet modern, components for the equipment. In order to achieve a high degree of non-dependability on any particular fuel, the station was to be suitable for burning natural gas, fuel oil, or coal. The high

proposed annual plant load factor (and the increased capital costs arising from provision for burning a variety of fuels) made a high thermal efficiency essential.

The project was further complicated by the fact that there was only a limited supply of cooling water available at the proposed site. This was one of the incentives which led to the consideration of a combination of a steam plant with a gas turbine to achieve a higher output than would have been possible with a pure steam plant using the same amount of cooling water—and also offering the possibility of raising the thermal efficiency at the same time.

A comparative study was made of a pure steam turbine plant and two different combined cycles employing gas turbines. The basic gas turbine cycle employed limited regenerative preheating of the compressor air, while the second alternative had no air preheating. At the site in question, with its limited supply of cooling water, the pure steam plant would have had a maximum output of only 64 MW, assuming a good efficiency and complete utilization of the average cooling water temperature of 8°C.

On an economic basis, the comparison made from the manufacturer's offers showed the cycle of Fig. 89.1 to be the most advantageous, and it was this that was finally used for the station.

The prime objective of the cycle is to use the gas turbine to increase the maximum power output of the station as much as possible above that of a pure steam plant while using the same amount of cooling water, and obtaining the

The MS. of this paper was received at the Institution on 19th February 1971.
* Kraftwerk Union AG, D8250 Erlangen 2, Werner-von-Siemens Str. 67, West Germany.

Table 89.1. Principal data

Items	Quantities
Gas turbine	
Output at generator terminals .	10·3 MW
Air inlet temperature . .	+10°C
Atmospheric pressure . .	760 mmHg
Total pressure ratio . . .	6
Speed of turbine . . .	4500 rev/min
Speed of generator . . .	3000 rev/min
Generator voltage . . .	6·3 kV
Natural gas fuel, H.C.V. . .	8600 kcal/Nm³
Steam plant	
Output at generator terminals .	{63·7 MW rated
	{68 MW maximum
Speed	3000 rev/min
Generator voltage . . .	10·5 kV
Steam conditions at boiler outlet .	190 kgf/cm² (gauge), 535°C
Reheat at 37 kgf/cm² (gauge) .	535°C
Feed-water temperature . .	297°C
Exhaust pressure . . .	0·023 kgf/cm² (abs.) at 8°C cooling water temperature

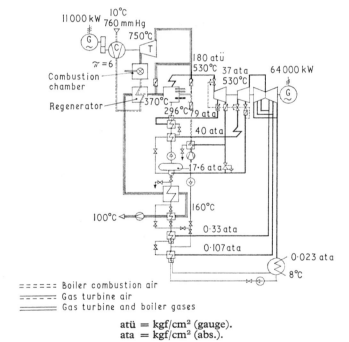

atü = kgf/cm² (gauge).
ata = kgf/cm² (abs.).

Fig. 89.1. Circuit diagram combined cycle

maximum possible overall efficiency under local conditions with the quality of fuel available. These requirements could best be satisfied by employing regenerative pre-heating of the compressor discharge air.

At the mean annual atmospheric conditions of 10°C and 760 mmHg, only 75·6 per cent of the turbine exhaust mass flow is supplied to the boiler burners for combustion of the fuel. This is the amount required to generate 196 t/h of steam when the turbine is producing 64 MW. The remaining 24·4 per cent of the turbine exhaust by-passes the boiler and takes no part in steam generation when the turbine is running at its most economical load.

There is a temperature difference of about 50 degC between the flue gas leaving the third pass of the boiler and the feed water entering the economizer at 296°C. The flue gases mix with the by-pass flow of exhaust from the gas turbine at 424°C. The mixed gas flow at 370°C passes to the regenerator.

During its passage through the compressor, the air is heated to 224°C. It is then raised in the regenerator to 342°C, at which temperature it enters the gas turbine combustion chamber. Approximately 36·6 per cent of the heat of the flue gas at 370°C is utilized in the regenerator. The remaining 63·4 per cent of the total flue gas heat in cooling down to stack temperature of 100°C is used for feed heating in a full-flow economizer situated between the l.p. and h.p. bled steam feed heaters.

The total amount of steam bled from the turbine in the combined cycle for heating the feed to 296°C corresponds approximately to that in a pure steam turbine of the same output, since in the combined cycle part of the feed, heating is carried out in the economizer. This arrangement makes it possible to operate with a cooling water flow of 10 200 t/h at 8°C to produce a pressure of 0·023 kgf/cm² (abs.) in the condenser. The shortage of cooling water is particularly marked during the winter months.

The excess gas turbine throughput relative to the combustion air requirements of the boiler burners at the most economical load results in a 16 per cent power gain over a pure steam plant. In addition, the high feed water temperature of 296°C, in conjunction with the regenerative preheating of the compressor discharge, results in a measured gross efficiency of 43·68 per cent on L.C.V., even though the stage of gas turbine development at that time permitted base-load operation with a temperature of only 750°C.

The results of the acceptance tests are shown in Fig. 89.2. The tests were independently supervised by Vienna Technical University. The graph shows the part-load

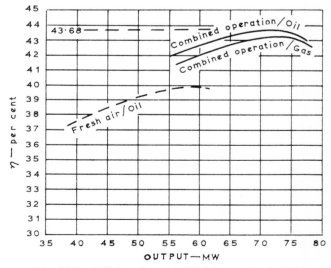

Fig. 89.2. Efficiencies on official test (on L.C.V.)

characteristics within the principal load range of the plant. The top curve refers to the burning of heavy oil in the boiler while the lower, thin curve refers to the burning of natural gas. The difference between the two curves is mainly because the heating surface of the reheater had to be designed for the use of the three different fuels. In all cases the gas turbine is operated on natural gas only.

The part-load characteristics of the steam plant when running alone were also measured and are shown by the dotted curve. The maximum output of the steam turbine under these conditions is still 60 MW if the combustion air from the forced-draught fan is heated in a steam pre-heater (Fig. 89.1).

MODES OF OPERATION

It was an important requirement that the risk of a complete disconnection of the station be reduced to a minimum. In particular, it was essential that the remaining generating unit would continue running in the event of trouble with the other.

During the course of operation to date, the following modes of operation have occurred.

(1) Emergency trip of the steam turbine from full load; the gas turbine continued running at full load.

(2) Emergency trip of the gas turbine from full load; the steam plant continued operating at full output.

(3) Shutdown of the boiler firing followed by an emergency trip of the steam turbine; the gas turbine continued operating at full output.

(4) Start-up of the boiler and solo running of the steam turbine without the gas turbine.

(5) Start-up and running of the gas turbine alone. An efficiency of 29 per cent resulted under these conditions from use of regenerative air preheating.

(6) Automatic shutdown of the gas turbine as a result of a failure of one of the two induced-draught fans. The steam plant remained in operation throughout the automatic change-over of the burners to fresh-air supply. The capacity of one induced-draught fan is adequate for this eventuality.

THE BOILER COMBUSTION AIR SYSTEM

Fig. 89.3 shows a cross-section through the boiler and turbine houses and the run of the exhaust ducting from the gas turbine to the boiler. The ducting runs vertically up the front of the boiler, branches into a Y-piece at the top, and passes to right and left of the boiler to the top of the air preheater in the fourth pass at the back of the boiler. The branch to the burners is just below the Y-piece and contains an isolating damper at this point (damper 1). Dampers (damper 2) to control the turbine exhaust flow, according to the firing of the boiler, are fitted where the by-passes enter at the back of the boiler (Fig. 89.4). These dampers regulate the pressure in the ducting, and in the boiler wind box when the damper (1) to the burners is open, to a constant 350 mm water gauge.

Another damper (3) is arranged at entry to the fourth pass between the steam-generating h.p. heating surface and the regenerator. This damper has an isolating function only and plays no part in the pressure control system. It need not be closed should the steam plant be shut down and the gas turbine continues running because a lower pressure exists at this point owing to the induced-draught fans, which will prevent the flow of turbine exhaust into the boiler.

Fig. 89.5 shows the positions of the dampers for the various modes of operation. When the gas turbine is running alone (top diagram), or when changing over from combined operation to solo operation of the gas turbine,

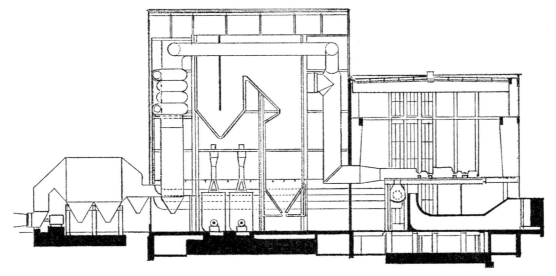

Fig. 89.3. Sectional elevation of boiler and turbine houses and gas turbo-set

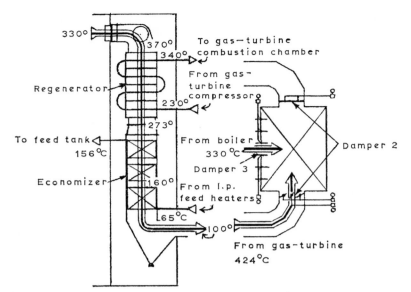

Fig. 89.4. Arrangement of fourth boiler pass

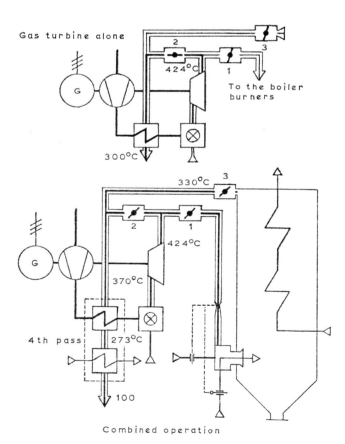

Fig. 89.5. Damper positions with gas turbine running alone and with the steam turbine

the full turbine exhaust is diverted through the by-pass lines.

The bottom diagram shows the positions of the by-pass dampers for combined operation. When a change-over is made from combined operation to solo operation of the steam plant, the forced-draught fan starts up automatically and supplies clean preheated fresh air direct to the boiler wind box. The by-pass control damper (2) closes automatically according to the diminishing exhaust supply from the gas turbine. The isolating damper to the wind box need not be closed automatically since the gas turbine takes a considerable time to come to rest, and so acts as a seal to prevent the escape of fresh air until the damper (1) is closed manually.

BOILER FIRING SYSTEM

Particular attention was paid to the design of the combined natural gas–heavy oil burners and the furnaces. Nine combined burners are arranged in three tiers facing the outside wall of the boiler house. Each burner is fitted with adjustable swirl vanes to regulate the flow of turbine exhaust or fresh air from the boiler wind box, according to the fuel flow rate.

The burners are also fitted with measuring sections (Fig. 89.5) to determine the flow rates of the fuel and combustion air to permit accurate adjustment of the fuel/air ratio. After pre-setting the calorific value of the fuel, the amount of combustion air is regulated automatically so that combustion of the fuel is practically stoichiometric with excess air values between 1·03 and 1·05. This control is also possible during simultaneous firing of natural gas and oil. The fuel/air proportional control system may appear somewhat complicated and expensive at first sight but, in fact, it has proved to be both economically and technically beneficial in practice.

The necessity for employing near-stoichiometric combustion lies in the low stack temperature of 100°C. The inlet temperature of the feed water is only 65°C, and hence the mean tube wall temperature at the end of the heating surfaces is between 68° and 70°C. In order to avoid low-temperature corrosion when burning sulphurous oil, the tendency to formation of SO_3 must be maintained as low as possible; low excess air achieves this.

In order to provide extra protection from sulphur corrosion for the cold end of the heating surfaces, an acid dew point recording device with indication in the central control room is provided. The inlet temperature of the feed to the economizer can be raised according to the reading of the dew point instrument by recirculating feed from the feed tank (Fig. 89.1). During operation to date, it has only seldom been necessary to make use of the system. Apart from these few cases, the plant has always been operated at the design final flue gas temperature of 100°C when burning both oil and natural gas.

After long periods of operation on natural gas the flue gas temperature has occasionally fallen below 100°C. Apparently, this is a result of a self-cleaning effect of the heating surfaces following operation of the boiler on coal or heavy oil. Although the natural gas contained no hydrogen sulphide, the flue gas temperature was maintained at 100°C by means of the recirculation system in order to avoid the temperature falling below the water dew point.

No corrosion damage to the cold end heating surfaces as a result of temperatures below the acid dew point, or the water dew point, has been observed during the course of the six years' life of the station. It is also certain that, in addition to stoichiometric combustion, the continuous mixing at 370°C of the turbine exhaust from burning natural gas free of sulphur with the flue gas from the boiler, and the resulting lowering of the SO_3 concentration, have been beneficial from the corrosion aspect.

No fouling of the heating surfaces as a direct result of using turbine exhaust for burning the fuel has been encountered. Fouling of the heating surfaces when firing the boiler on oil was also negligible. However, in the past,

when burning coal simultaneously with oil or natural gas, slagging has occurred in the furnaces; but not enough to make cleaning of the heating surfaces necessary during the annual running time between overhauls. The h.p. heating surfaces of the boiler have not suffered any corrosion damage through the burning of oil.

A further problem which arose during the planning stage concerned the question of complete combustion and flame lengths when using turbine exhaust gas as the combustion air. The composition of the turbine exhaust is: O_2 content, 17·6% by volume; CO_2, 1·6%; H_2O, 3·3%; CH_4, 0·1%; and N_2, 77·4%. The exhaust contains no SO_2.

In the absence of detailed practical or theoretical data on the subject, it was thought that the burner flame when using turbine exhaust would require a greater distance to achieve complete combustion than would be the case when using fresh air. In actual fact, the result was exactly the opposite.

The shapes of the flames are shown in Fig. 89.6. When burning oil in fresh air the flame is long and narrow and stretches practically the whole length of the furnace. When burning oil in turbine exhaust, on the other hand, it is much shorter and wider. This fact, which naturally is also affected by the type of atomizing burner with automatic oil-viscosity control used, can be explained in the following manner.

The temperature of the turbine exhaust at 420°C is 170 degC higher than the 250°C of the preheated fresh air. As a result of the higher temperature, the droplets of the oil spray produced by the steam-atomizing burner evaporate more quickly than when using fresh air.

Two further factors also have an effect. The mass flow of the turbine exhaust required for the complete combustion of a unit quantity of oil is slightly more than 20 per cent greater than the amount of fresh air required, assuming the same value of excess air between 1·03 and 1·05 in both cases. Allied to this is the greater specific volume of the turbine exhaust at the higher temperature, as opposed to the lower temperature fresh air (Fig. 89.6). Both factors

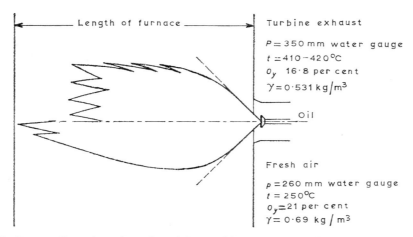

Fig. 89.6. Burner flame lengths using either turbine exhaust or fresh air for combustion

lead to an increase in the velocity of the turbine exhaust passing into the combustion zone compared to air, with the result that both the swirl in the primary atomizing zone in front of the burner tip and the turbulence in the secondary combustion zone are notably more intensive than when using fresh air. The increased turbulence in the secondary combustion zone is particularly effective in furnishing the remaining oil at the end of the flame with oxygen.

In comparison with these results, the combustion of natural gas, using turbine exhaust as the combustion air, exhibits hardly any difference in flame shape from that produced when fresh air is used, since in this case the mixing phase with the steam is eliminated and thus complete combustion takes place relatively more rapidly.

OUTAGES AND DEFECTS

Forced outages since 1965 amount to 635 h for the gas turbine and 1392 h for the steam plant. These figures include 194 h when both turbo-sets failed together. Detailed figures, with the outage hours expressed as a percentage of the operating hours, are given in Fig. 89.7.

When reliability is defined as:

$$\frac{\text{total operating hours}}{\text{total operating hours} + \text{forced outage hours}}$$

the first six years' operation results in a mean figure of 98·5 per cent for the gas turbine and 96·8 per cent for the steam turbine.

The causes of the various outages which have affected the station cannot be described fully in this report, therefore only a brief summary will be given.

It should be stated at the outset that the components and sections of the plant which provide the interconnection between the gas turbine and steam turbine cycles have never given rise to any outages. Therefore, when the steam turbine has failed for some reason, the gas turbine has almost always remained in operation, and vice versa. On most occasions when the whole station has been shut down the fault has been in the supply system.

The gas turbine has most often been tripped owing to the supply system frequency rising above 51 Hz. Other causes, such as tripping initiated by the flame monitor, low gas supply, failures to the station auxiliary supply system, etc., have already been described elsewhere (I)*.

One outage of 400 h that affected the steam turbine plant alone was caused by flood damage to the cooling water intakes. Otherwise, all damage or defects have been of a minor nature and without exception were discovered only after the plant had been shut down for a planned inspection.

Damage which affected the gas turbine specifically was, first, a crack in the turbine inner casing which carries hot gas on the inside and compressed air on the outside; the crack was welded over. Second, during the 1967 overhaul it was noticed that a rotor seal ring showed permanent deformation. After investigation this was attributed to insufficient cooling. An increase in the cooling air supply brought an improvement because at the next overhaul the new ring exhibited no deformation.

Neither the combustion chamber nor the burners have required any replacements so far. Other spare parts replaced have been confined to wearing components, gaskets, seals, bolts, screws, etc.

No trace of erosion damage has been found either in the compressor blading or the turbine blading. Hence, it

* *References are given in Appendix 89.1.*

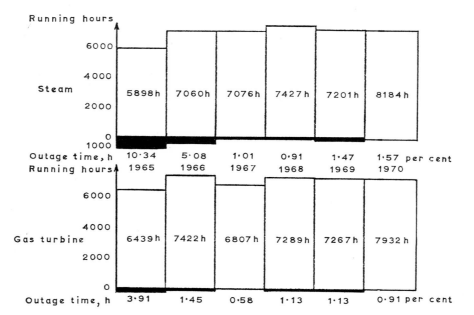

Fig. 89.7. Annual operating and outage times of the gas turbine and steam turbine, 1965–70

would appear that the fabric air-intake filter functions satisfactorily. After 30 000 h the filter showed an increased carry-through caused by deterioration of the filter medium (changed during the 1969 overhaul). It resulted in some fouling of the compressor and caused a slight reduction in power output.

The electricity-supply situation in Austria is such that in summer almost the whole load can be supplied from hydroelectric stations. Consequently, the Hohe Wand station (2) (3) is not required and is shut down, normally between July and September. Planned inspections are carried out during this period. In the life of the station so far, inspections have been carried out annually, i.e. normally after around 7000 h running. From 1969 the gas turbine will be opened up for a major inspection only every two years, normally after 14 000–15 000 h running.

GENERAL RUNNING CHARACTERISTICS

The running smoothness of the turbo-alternators is extraordinarily good. The amplitude of the vibration at the bearings of the steam turbine is between 2 and 5 μm and at the gas turbine bearings between 3·5 and 8·5 μm.

With regard to noise nuisance in the surroundings, except for the first few months during testing, no complaints have been received. The noise level in the plant itself is within normal figures—near the gas turbine 96 dBA and near the boiler burners 88 dBA. The boiler house is distinctly quieter than the turbine house because there is no compressor noise.

COSTS

At the time of completion in 1965 the specific capital cost was 4500 Schillings per installed kilowatt (£72 when £1 = 62·4 Schillings). This figure includes all miscellaneous costs, such as site acquisition, connecting services, construction of access roads, fencing, interest charges, site supervision costs, planning costs, central administration costs, costs of official tests and approvals, and for the services of independent observers and consultants.

It should be remembered when evaluating these costs that the station is equipped to burn three different types of fuel. This results in relatively high expenditure on facilities for the storage and transport of the coal, oil, and natural gas.

Running costs for fuel, maintenance, and overhauls were 0·14 Schilling/kWh in 1970 (about 0·54d./kWh when £1 = 62·4 Schillings). The mean annual net efficiency calculated from the fuel consumption and the useful power supplied to the system is about 40 per cent.

APPENDIX 89.1
REFERENCES

(1) LAHR, G. 'Die Verwendung von Erdgas in einem kombinierten Gas-Dampf-Prozess', Gas-Wärme Int. 1967 **16** (No. 7), 378.
(2) GOEBEL, K. ' "Hohe Wand" gas turbine–steam turbine power station', Siemens Rev. 1966 **XXXIII** (No. 5, May), 262.
(3) CERMAK, H. 'Betriebserfahrungen mit der Kombination eines Gasturbinen-Dampfkraftprozesses in der Anlage "Wärmekraftwerk Hohe Wand" ', Mitt. Verein. Grosskesselbetrieber 1970 **50** (Bk 3, June), 161.

C97/71

EXPERIENCE WITH NATURAL GAS
FIRING IN LARGE BOILERS

C. HIJSZELER*　　K. VAN DUINEN*

The firing of natural gas in large steam boilers has increased greatly in Holland since Groningen natural gas was made available to Dutch industry. A survey is given of boilers in larger industrial plants that use natural gas as a fuel. The gas supply system is designed for maximum security of gas delivery at reasonable cost. To ensure operational safety of gas fired plants, regulations that must be followed have been drafted for gas fired installations. Fuel supply systems that are considered safe are described. The practical experience of firing natural gas in boilers given in this paper is centred around the specific combustion characteristics of natural gas. Both burner and boiler design criteria are discussed.

Natural gas makes possible the use of a small number of large-size burners, due to the wide turn-down ratios realized with natural gas. Experimental work in test furnaces is required to achieve, with a minimum of delay, optimum designs. Special attention has to be given to the tendency of burners to generate noise and to give rise to pulsation problems.

To obtain optimum results of gas firing the boiler design should be adjusted to natural gas firing. The radiation characteristics of a gas flame are different from those of pulverized coal and oil flames, as well as the flue gas rate. Results are shown for conversion jobs and new designs. The absence of sulphur in Dutch natural gas makes it possible to drop the exhaust temperature of the boiler below the temperature of oil fired boilers.

INTRODUCTION

SINCE THE INTRODUCTION of Groningen gas to the Dutch energy market a great expansion has taken place. This is true for both the domestic and the industrial markets. Fig. 97.1 illustrates the amount of gas sold in some market areas. As can be seen, the introduction of natural gas in the industrial market has been swift.

In Holland about 70 per cent of the energy used in industry is produced in boilers. In order to obtain a high rate of gas sales in industry at an early stage, attention was given to the possibilities of using gas as an under-boiler fuel. The price level of the gas in Holland and technical–commercial efforts were adjusted to allow achievement of the target: penetration in the boiler fuel market.

SURVEY OF BOILER INSTALLATIONS
IN DUTCH INDUSTRY

As the subject of this paper is the experience of the use of natural gas in large boilers, the survey has been restricted to boiler installations used in the larger industrial enterprises, i.e. industries using more than 2×10^6 m³ gas/ annum. Furthermore, this restriction is useful because these industries normally have a higher fuel supply pressure (3–8 bar) than the smaller industries. This fact affects

The MS. of this paper was received at the Institution on 8th January 1971 and accepted for publication on 8th March 1971. 33
★ N.V. Nederlandse Gasunie, Groningen, Netherlands.

the design of the heat gas supply system and that of the burner.

In the survey given in Table 97.1, subdivisions have been made according to the type and capacity of the boiler, the type of fuels used, and the original design of the boiler—original, that is, for coal, oil, or for natural gas. As can be seen from this survey, there are an appreciable number of installations that use natural gas exclusively as a fuel. This is only possible when the gas supply is reliable. The gas transport system has to be designed with this safe supply situation in mind.

The advantage of using a single fuel, i.e. natural gas, for a boiler is that the design of the boiler can be optimized for this fuel. This means making full use of the favourable characteristics of natural gas.

Dual-fuel installations are used only in those cases where the gas supply will be interrupted intentionally, e.g. because of gas transport economy. Interruption of the gas supply takes place in winter time at moments of high load on the transport grid; the cause being high gas consumption to meet heating demands in the domestic market.

THE GAS SUPPLY SYSTEM

The majority of the gas reserves in Holland are situated in the northern part of the country. Fig. 97.2 gives an impression of the main transport system operating at a maximum pressure of 40 bar. From this system the supply

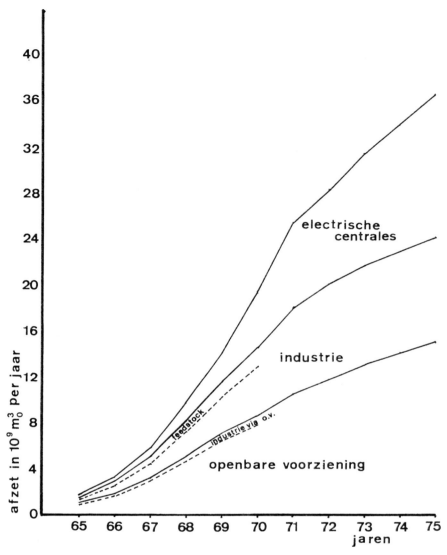

Afzet in 10^9 m_0^3 per jaar = Sales in 10^9 m_0^3 per year.
Electrische centrales = Central electricity.
Industrie via o.v. = Industry via public distribution.
Openbare voorziening = Public distribution.
Jaren = Years.

Fig. 97.1. Sales of gas during the period 1965–75 divided into the different buyer–consumer categories

lines branch off to individual consumers, industry, power stations, and municipal distributors.

The gas transport system is compact because of the size of the country and the concentrated high-energy consumption. This enables us to safeguard the gas supply by using loops in the system without unduly high transport grid investments.

In addition, for the same reason the load factor of the grid does not influence the selling price of the gas to the extent occurring in other countries. Both factors, safe supply and the independence of gas price to load factor,

stimulate the utilization of gas in industry without consideration of alternative fuels.

The gas is supplied to consumers through a measuring and regulation station. Supply pressures are normally between 3 and 8 bar. Preference is given to 8 bar.

The gas technical equipment, owned by the gas transport company, always includes a stand-by unit. Should there be a failure in the supply system, the stand-by unit comes into operation automatically without any interruption of the gas supply. Fig. 97.3 illustrates the system.

The quality of the natural gas delivered to users is of a

Table 97.1. Water-tube boilers in the Netherlands

Type	Industrial	Power stations	Boilers and conversions at February 1971, Total
Number of new units			
Gas fired . . .	59	5	64
Oil and gas fired . .	7	22	29
Number of conversions			
To natural gas:			
from coal+oil . .	—	3	3
from oil . . .	73	—	73
from coal . . .	16	34	50
To oil+gas:			
From oil . . .	15	8	23
From coal+oil . .	—	5	5
To coal+gas:			
from coal . . .	3	4	7
from coal+oil . .	—	3	3
Number of boilers			
Gas fired . . .	148	42	190
Dual firing . . .	25	42	67

very constant quality. No provisions are made in measuring and regulation stations to check the quality or to regulate the adjustment of gas-using equipment.

SAFETY OF GAS UTILIZATION

The safe use of gas is the next important item after security of supply. Safety regulations have been formulated by representatives of the manufacturers of equipment, equipment users, government, and gas suppliers.

These safety regulations concern the design and construction of the gas transport lines in the terrain and premises of the users. The aim is to exclude the possibility of unwanted fires and explosions, and to ensure proper operation of the gas-firing systems. In order to achieve this, the gas supply system design for the gas-using equipment is thoroughly checked. After installation is completed, strength and tightness tests are carried out.

Furthermore, the equipment used in conjunction with the burner proper and the procedure of utilization of the burner have to comply with certain regulations. In Fig. 97.4 the safety system used for a single burner installation is given.

Explosions in the combustion space can only occur if a certain amount of combustible gas–air mixture is present therein. This can occur before starting up as a result of gas leakage through a valve. Therefore double-stop and bleed valves are required.

In order to make sure that before start-up of the burner the combustion space contains only air, a purging period has been prescribed. During the start-up the auxiliary or the main burner might fail to ignite, or during operation the flame might extinguish.

The time during which a gas–air mixture is allowed to enter the combustion space without a signal from the flame-sensing device, demonstrating the presence of a flame, is restricted. This limits the amount of combustible mixture present at any given time in the combustion space.

It is essential for the safe operation of a gas fired installation that a stable flame is present even under adverse firing conditions. Under fuel-rich conditions this will ensure that a maximum amount of fuel is burned, producing inert gases, such as CO_2 and H_2O, that will bring the mixture of combustible gases present and additional combustion air entering the combustion space, e.g. after fuel cut-off or readjustment of the burner, outside the explosive limits. Fuel-rich mixtures occur as a result of a lack of air. A minimum air pressure switch, or equivalent equipment, that cuts the fuel supply must be incorporated. The above-mentioned fuel-rich operating conditions can be caused by the occurrence of higher gas pressures than the burner design allows. A maximum gas pressure switch that stops the gas supply has to be installed. Too low gas pressures endanger the stable operation of burners. The safety system therefore must contain a minimum gas pressure cut-off.

BURNER CONSTRUCTIONS

Burners for large boiler installations in Holland are mostly designed for a gas supply pressure before regulation of about 8 bar. The pressure before the gas ports ranges between 0·5 and 6 bar. The main advantage of using high gas pressure is that the cost of gas piping and valves is low in comparison with low-pressure installations. Smaller size equipment results in better accessibility and easier handling of equipment.

If the burner is not specifically designed for the higher gas pressures, noise problems might arise. On the other

---

Interpretation of key on Fig. 97.2.

Hoofdtransportnet—per 31 december 1969 voltooide projecten = Main distribution centres—completed projects as at 31st December 1969.
Hoofdtransportnet—bouwprogramma 1970 = Main distribution centres—1970 building programme.
Voedingsleiding—per 31 december 1969 voltooide projecten = Domestic services—completed projects as at 31st December 1969.
Voedingsleiding—bouwprogramma 1970 = Domestic services—1970 building programme.
Regionaal transportnet—per 31 december 1969 voltooide of in aanleg zijnde projecten = Regional transport centres—completed projects or projects still under construction as at 31st December 1969.
Regionaal transportnet—toekomstige leidingen = Regional transport centres—future consumers/buyers.
Voedingsstation aardgas = Natural gas supply stations.
Meet-en-regelstation = Pressure regulating and metering stations.
Toekomstig meet-en-regelstation = Future pressure regulating and metering stations.
Compressorstation = Compressor stations.
Toekomstig compressorstation = Future compressor stations.
Exportstation = Export stations.

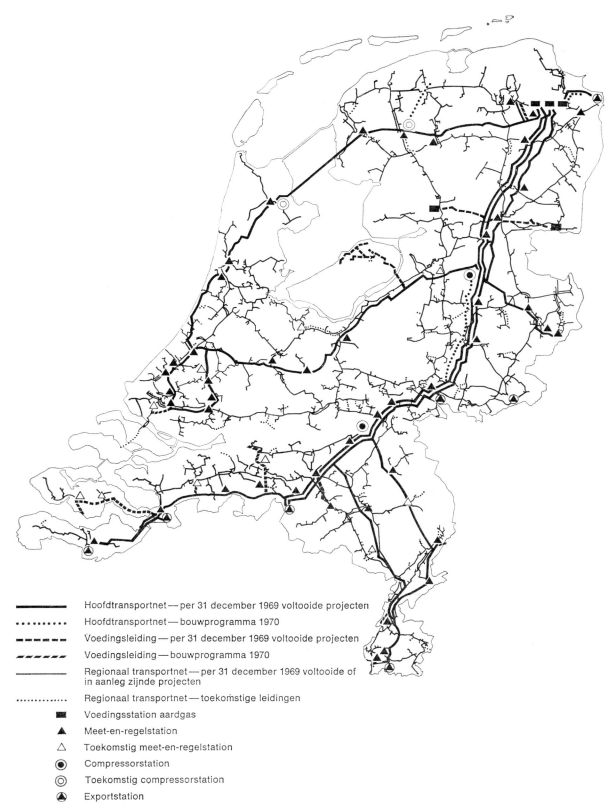

Hoofdtransportnet — per 31 december 1969 voltooide projecten

Hoofdtransportnet — bouwprogramma 1970

Voedingsleiding — per 31 december 1969 voltooide projecten

Voedingsleiding — bouwprogramma 1970

Regionaal transportnet — per 31 december 1969 voltooide of in aanleg zijnde projecten

Regionaal transportnet — toekomstige leidingen

Voedingsstation aardgas

Meet-en-regelstation

Toekomstig meet-en-regelstation

Compressorstation

Toekomstig compressorstation

Exportstation

Fig. 97.2. Main transport system operating at maximum pressure of 40 bar

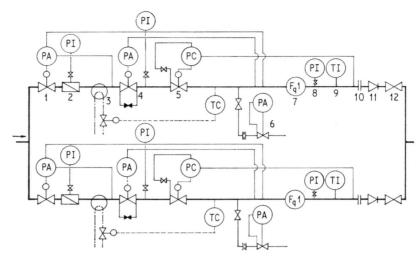

1. Safety cut-off valve.
2. Gas filter.
3. Gas heater.
4. Auxiliary pressure regulator.
5. Main pressure regulator.
6. Safety blow-off.

7. Flow meter.
8. Pressure gauge.
9. Thermometer.
10. Restriction orifice.
11. Non-return valve.
12. Valve.

Fig. 97.3. Supply system

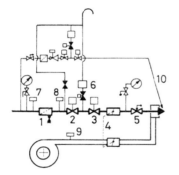

1. Strainer.
2. Automatic safety valve.
3. Automatic safety valve.
4. Air and gas flow control.
5. Hand-operated valve.
6. Bleed valve with flow indicator.
7. Maximum pressure switch.
8. Minimum pressure switch.
9. Minimum pressure switch.
10. Ignition burner gas supply.

Fig. 97.4. Safeguarding system for a single installation

hand, our experience has shown that high gas pressures can be used for burners without giving rise to excessive noise. In many cases noise problems are caused by instability of combustion. Another possible cause is that the distribution of the fuel gas over the total air volume is such that combustion takes place where unstable flow or combustion air temperatures occur. This results in an uneven combustion, which gives rise to pressure pulsations and noise.

Although it is generally accepted that gaseous fuels burn easier than other fuels, it must be emphasized that the combustion of natural gas mainly containing methane can cause major problems in comparison with the combustion of other gases containing free hydrogen or appreciable amounts of higher hydrocarbons. Under such circumstances, production of noise, as mentioned above, or insufficient stability create problems.

There are three factors that together govern the stability of the burner: time, temperature, and turbulence. The time factor is mainly given by the speed of combustion of the gas or gas–air mixture at the point of ignition in relation to the local flow velocity. If the flow velocity exceeds the combustion speed, the flame tends to be unstable. The maximum allowable speed at the exit of the gas at the point at which the combustion has to stabilize is given by the gas analysis.

The temperature of the gas or gas–air mixture at the point of ignition influences the combustion speed. Higher temperatures result in higher combustion speeds. Frequently, constructions are used that increase this temperature. Possibilities are the use of auxiliary burners that heat the gas at the exit of the main burner, or the use of constructions that cause recirculation of combusted gases to the flame foot. Examples are those producing swirl in the combustion air stream, resulting in a low pressure in the centre of the flame, thus causing reverse flow of burned gases. Fig. 97.5 gives diagrams of some burners.

Priority has to be given to stabilizing methods that guarantee stable combustion as soon as possible after a cold start. In this connection the use of ceramic materials as stabilizing agents has to be discouraged.

The intensity of the mixing of gas and combustion air in industrial burners (turbulent mixing occurs) increases

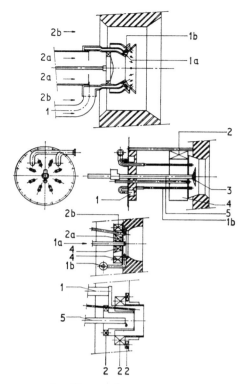

1.   Gas supply.
1a. Auxiliary gas supply.
1b. Main gas supply.
2.   Air supply.
2a. Auxiliary air supply.
2b. Main air supply.
3.   Stabilization disc.
4.   Air registers.
5.   Oil gun.

Fig. 97.5. Diagrams of burners

the combustion speed. In some instances a certain pre-mixing of air and gas is used to improve burner stability. Stability of combustion not only has a bearing on noise and, as mentioned, on safety of operation, but determines the capacity which can be reached by a certain burner size.

Both fuel and air being gaseous, allowing simple flow measurement and proportioning together with the ease of combustion of gas as a fuel, enables us to obtain very low percentages of maximum load as a minimum load. The minimum load is normally not given by the possibilities of proper combustion but by the accuracy and reliability of the flow proportioning system.

The big turn-down ratio that can be reached, normally 10:1, makes it possible to use a minimum number of very large capacity burners. This reduces the price of the installation.

BURNER DEVELOPMENT

Since the introduction of natural gas in Holland a considerable amount of gas burner development work has been done. This work incorporated experimental work that had been done on open test beds and in existing boiler

installations. A test installation has been available for almost two years for burner development work. This test furnace (Fig. 97.6) is situated at the Gasunie laboratories, and has been installed by four major Dutch burner manufacturers in co-operation with Gasunie. By paying an annual fee, the burner manufacturers are entitled to use the installations for a certain period. The maximum capacity of burners that can be tested is 50×10^6 kcal/h. The combustion space is 8 m long, 4 m wide, and 5 m high.

CONSEQUENCES OF GAS FIRING ON BOILER DESIGN

In the design of fossil fuelled boilers, the fuel characteristics are of paramount importance and impose various limitations. Compared with coal and liquid fuel, the most outstanding features of natural gas are:

(1) complete combustion with no unburnt particles;
(2) no ash, therefore no ash disposal is necessary;
(3) cleanliness of the flue gas;
(4) low dew point of the flue gas (only with sulphur-free natural gas, as with Slochteren natural gas);
(5) no air pollution from ash, soot, or sulphurous gases;
(6) low emissivity of the flame as no solid particles are present; and
(7) reduced maintenance of boiler plant.

A number of conclusions can easily be drawn from these fuel characteristics:

(1) A furnace hopper is not necessary as there is no ash to dispose of.
(2) Equipment for cleaning the heating surface, such as soot-blowers and shot cleaning, can be omitted.
(3) There are no limitations on the furnace exit gas temperature from the fuel side as there are no ash fusion problems.
(4) There are no limitations on tube pitching or flue gas velocity as there is no danger of erosion or fouling. An extended heating surface can be used to obtain a compact design.

Table 97.2 summarizes a few typical parameters in combustion and furnace performance of natural gas compared with fuel oil and coal.

The interesting thing in this table is not the absolute value of the parameters—F.E.G.T. of 1290°C would not be acceptable for most grades of coal—but their ratio for the different fuels. Compared with natural gas, fuel oil has a lower F.E.G.T. as well as a lower specific flue gas quantity. Coal, on the other hand, has about the same F.E.G.T. while the specific flue gas quantity is larger. This strongly influences the design of superheaters—or the performance of superheaters in multi-fuel installations—and we will revert to this subject when we discuss boiler conversions.

Sulphur-free natural gas compares very favourably with

Fig. 97.6. Test furnace at Gasunie laboratories for work on burner development

sulphur-containing heavy fuel oil. In the case of an economizer as the last heating surface, the feed water inlet temperature should be in the range 140–150°C to avoid acid dew point corrosion. In industrial boiler installations, feed water is normally supplied from a deaerator operating at 105°C, bringing about the application of some form of feed water preheating. In the case of an air heater as the last heating surface, the air is usually preheated to about 80–100°C as a result of the water vapour content. This means that the stack temperature can be about 40 degC lower, resulting in an increase in boiler efficiency of about 1·5 per cent.

Table 97.2. Combustion and furnace performances

Parameter	Coal on grate	Pulverized fuel	Fuel oil	Natural gas
Lower heating value	6300 kcal/kg	6300 kcal/kg	9600 kcal/kg	7560 kcal/N m³
Excess air	30 per cent	20 per cent	5 per cent	5 per cent
Flue gas quantity per 1×10^6 kcal heat input .	1487 N m³ 1986 kg	1375 N m³ 1843 kg	1207 N m³ 1589 kg	1298 N m³ 1612 kg
Furnace exit gas temperature in the same furnace at the same heat input	1270°C	1290°C	1250°C	1300°C

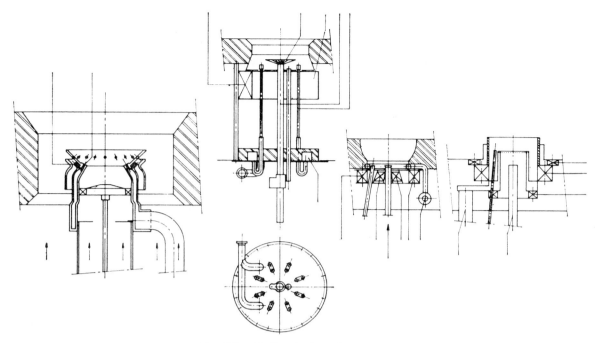

Fig. 97.7. Boiler designed to use gas

BOILERS SPECIFICALLY DESIGNED FOR GAS FIRING

In recent years a market has developed in Holland for industrial boilers operating on gas. Fig. 97.7 shows a boiler designed to take full advantage of this fuel. This type of boiler is designed to have a steam capacity between 25 and 100 t/h, and is equipped with one or two burners.

In what respects does this boiler design differ from other boiler designs? First, the furnace dimensions are much smaller than in the usual type of oil fired boilers. The furnace loading can go as high as 1 000 000 kcal/m³ h, giving a residence time of about 0·4 s. The time required to achieve complete combustion is only about 0·25 s for natural gas fired in forced draught burner installations, being two or three times less than with oil firing.

It seemed logical to install the burner(s) at the bottom of the furnace. In outdoor installations, most frequently encountered in process industries, the burner and associated equipment are protected against the weather to a certain extent.

The heating surface heat load in the furnace, expressed as the total heat input divided by the projected wall surface, is 500 000–600 000 kcal/m² h, giving a furnace exit temperature of over 1500°C. However, owing to the low emissivity of a natural gas flame, the heat absorption by the furnace walls is no higher than with oil firing. After passing over a convection evaporator with gas velocities up to 25 m/s, the flue gases are cooled down in a finned tube economizer to a stack temperature of 175°C with a feed water temperature of 105°C. This boiler can be assembled in the workshop and shipped to its destination in one piece, with the exception of the economizer which will be shipped as a separate part. In comparison with a normal two-drum packaged boiler (designed for oil), the boiler weight is reduced by 30 per cent and its volume by 20 per cent at a price of 80–85 per cent of its oil fired counterpart.

CONVERSION OF BOILERS TO GAS FIRING

With the penetration of natural gas in the industrial energy market, the impact of this new fuel is not limited to new boilers. Undoubtedly, in Britain, as in Holland, quite a number of existing boilers will be converted to natural gas firing. The way in which this is done, and the points to which attention must be paid, depend on the existing fuel and firing equipment. In our experience three cases have to be considered separately, according to the existing firing system: (1) coal fired on a grate; (2) coal fired in burners as pulverized fuel; and (3) oil fired.

Coal fired on a grate

In this case the most appropriate place to install gas burners is at the bottom of the furnace that becomes free after removal of the grate. No alteration need be made to the pressure parts of the furnace. In Fig. 97.8 an example of such a conversion is shown on three identical boilers, each with a steam production of 100 t/h at 36 atm and 450°C. Four burners were installed in a common wind box. The heat absorption by the superheater unexpectedly turned out to be higher than with coal firing, and could easily be corrected by means of the gas by-pass over the superheater. Measurements taken showed that the furnace exit gas temperature was considerably higher than expected.

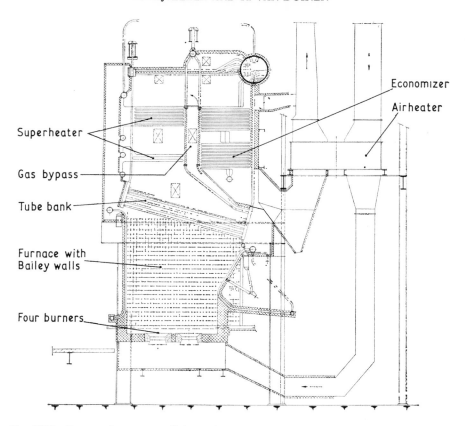

Fig. 97.8. Conversion to gas firing of boiler which had used coal burnt in a grate

In our opinion this must be attributed to the location of the burners in this furnace, whereby only part of its volume is swept with hot radiating gas. Both forced draught and induced draught fans normally pose no problems as the value of combustion air and flue gases is considerably lower as a result of the decrease in excess air—about 20 per cent decrease in this case.

Coal fired in burners

Although no operating experience is yet available of this type of conversion, a number of W-flame boilers are in hand for conversion. Burners are installed in the front or side walls according to layout desirability. The height of the burner level is chosen so that it will give the same superheater performance. If this leads to an unnecessarily high burner level, flue gas recirculation is applied to attain the same range for full superheat temperature.

Oil fired

Fig. 97.9 shows an industrial boiler that was converted from oil firing to gas firing. Steam capacity is 70 t/h at 73 atm and 510°C. Eight oil burners were replaced by three gas burners. Apart from the new tube bending in the front wall for the burners, the only modification of the boiler consisted in cutting out one loop of the superheater,

as can be seen in the drawing. The gas by-pass across part of the primary superheater was not sufficient to cope with the increased absorption by the superheater.

In a number of cases it was found necessary to decrease the steam temperature by about 10 degC to avoid superheater tube metal temperatures in excess of their design limit for oil firing.

On all jobs stack temperature did not differ markedly from the stack temperature before conversion. The furnace exit gas temperature with gas firing is usually higher, but the heat absorption by the convection part of the boiler compensates for this as there is no reduction in heat transfer by fouling.

If the feed water temperature at the inlet of the economizer is decreased from 140°C to 105°C, as is advisable on conversion from oil firing to gas firing, the stack

Table 97.3. Breakdown of conversion costs

Equipment						A	B
Boiler .	.	.	.	.	.	10	90
Gas firing equipment .	.	.	.	.	50	50
Control system .	.	.	.	.	.	10	30
Safeguarding system .	.	.	.	.	30	30
Total .	.	.	.	.	.	100	200

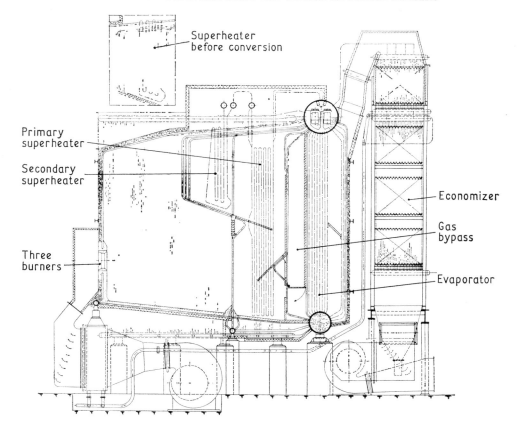

Fig. 97.9. Conversion to gas firing from oil firing

temperature will decrease by about the same amount. This will result in an interesting improvement in the overall boiler efficiency.

COSTS OF CONVERSION

Regarding the breakdown of the costs associated with the conversion of boiler plant to gas firing, Table 97.3 can serve as a rough guide. If we take the total conversion costs of an oil fired boiler with automatic controls as 100, the division is indicated in column A. In column B the same is given for a boiler with coal firing on a grate. The costs of the gas firing equipment and the safeguarding systems are equal in both cases, as could be expected. The combustion controls of an automatic oil fired boiler need only minor adjustments. The main difference between case A and case B is the boiler itself, since in case B a new furnace floor has to be constructed.

CONCLUSIONS

This paper gives some criteria for gas fired boiler installations. Though gaseous fuels are easily burnt, the design of a gas burner for natural gas requires an appreciable amount of special knowledge and experience in order to arrive at optimum results.

If natural gas is the only fuel to be used, the boiler can be designed specially for gas. In this case the minimum boiler size and the maximum possible efficiency can be obtained.

The installations using natural gas as a fuel can be safe, provided that proper regulations for the installation are followed.

The experience in the Netherlands indicates that the users of gas fired installations—converted as well as newly built—are satisfied with the performance of their installations. It compares favourably with the installations using other types of fuel.

Discussion

F. A. Allan Arnhem

The authors of Paper 97 mentioned the conversion of coal-fired boilers to gas firing. The Provinciale Gelderse Elektriciteits Maatschappij reports the conversion of two 400 t/h boilers in 1970.

Our company decided on a conversion to gas, combined with oil firing. The reason was that, for relative low cost (15 per cent of the total conversion because the oil-transport installation was already in existence) the possibility of oil firing gave us the opportunity to conclude a favourable contract with Gasunie (interruptible gas).

The two boilers converted were of the Sulzer type (400 t/h, 540°C–540°C, 190 kg/cm²), suited for firing low-volatile coal and heavy oil (Fig. D1).

The original construction consisted of 24 combined coal–oil burners in the top of the W. The boilers were built in 1962/1963. Instead of these 24 combined coal–oil burners, six combined gas–oil burners were installed, three in the vertical front wall and three in the back wall opposite to the first three. Fig. D2 shows the front wall inside the boiler. The maximum burner capacity for oil is 5000 kg/h and for gas 6500 Nm³/h (standard cubic metre). This Dutch-made air atomized burner is shown at the top of Fig. 97.5. The range of control for oil as well as for gas amounts to 1/7. The guaranteed percentage of oxygen amounts to 0·6 per cent for oil (load 80–100 per cent) and to 1 per cent for gas (40–100 per cent).

The study of this conversion showed that, especially with oil firing, the evaporator furnace would absorb too much heat. It might have been possible to use flue-gas recirculation, but it appeared that the cheapest solution was to cover part of the furnace with a heat-insulated layer of mouldable refractory material and to use hand-tilting burners. These burners are aligned down (maximum 25°) for gas firing and aligned up (maximum 25°) for oil firing. (Fig. D3 shows two burners tilted in different ways.) Further alteration of the heating surfaces was not necessary. The main order for this conversion was given on 1st December 1969. The first boiler came into operation on 15th July 1970 and the second boiler on 15th October 1970. The time of non-availability was nine weeks for the first and eight weeks for the second boiler.

The percentage breakdown of conversion costs (see Table 97.3) was:

Boiler	30
Gas-firing equipment	50
Control system	9
Safeguarding system	11
	———
	100%

The total cost per tonne of steam was 3750 Dutch guilders.

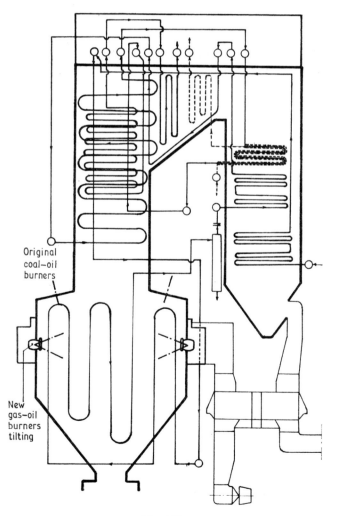

Original coal–oil burners

New gas–oil burners tilting

Fig. D1

N. Gasparovic Delft

The comparison of both groups of combined cycles, namely the high-efficiency cycle and the recuperation

Fig. D2

Fig. D3

cycle, as shown in Fig. 76.7 of Paper 76, is original and very instructive.

First of all I should like to discuss the 'apparent anomaly', that the success of combined-cycle power plants has not led to repeat orders in the U.S.A. I would add that it has not led to repeat orders in other countries, to any significant extent. The author gives us at least two reasons: availability of natural gas and the impact of nuclear power.

Let us see if there are any other reasons. What I am going to say applies only to high-efficiency combined cycles for base-load generation.

If the gas-turbine and the steam-turbine plant are properly matched the estimated gain of 5 per cent in heat rate is correct. The extraordinarily good gross efficiency of the Hohe Wand station must be partly due to the low air-inlet temperature of 10°C and the low cooling-water temperature of 8°C.

The author of Paper 76 regards the potential reduction of capital costs of about 5 per cent to be a significant claim. I am afraid that it will remain only a claim if we make the following comparison.

The total power output of the combined plant has to be divided as between 85 per cent from the steam plant and 15 per cent from the gas turbine. On the assumption that the installed cost per kilowatt of the gas turbine is 30 per cent below that of a large steam plant (the building costs being ignored) the absolute gain becomes 4·5 per cent. But on the other hand we have to consider that the steam plant becomes distinctly more expensive (say 6 per cent) than a conventional steam plant of the same total power output. This 6 per cent follows because of the reduction in size, because of the additional ducts and dampers, and because of the higher temperature at the firebox inlet. This means, with regard to the fractional output of the steam side of the plant, that there is an increase of specific cost of 5·1 per cent. So the combined power plant gives a net overall cost increase of 0·6 per cent in place of the author's 5 per cent decrease in comparison with a conventional steam plant of the same size.

My point thus is, that it is not permissible to compare a conventional steam power plant of 85 per cent power output with a combined cycle plant of a higher output, namely 100 per cent. A valid comparison can only be made if the two plants have the same output.

My analysis shows that the combined-cycle plant of the high-efficiency type is more expensive. Even if my rough estimate is wrong, I suggest that the difference in investment costs in favour of the combined cycle can be only marginal.

The companies involved have to consider not only the heat rates and investment costs but also the maintenance and personnel costs which are higher for the combined-cycle plant. The maintenance schedule is tighter than in a power plant with only one thermodynamic cycle. The availability (an aspect of great importance) for the combined power plant operating at full load is lower.

Additionally, to have to handle two or even three different fuels cannot be considered an advantage.

To sum up, under normal conditions, there is no significant gain from the use of a high-efficiency combined cycle. Naturally, there always can exist some special circumstances which can favour a combined-cycle plant, as for example at Hohe Wand where the amount of cooling water available restricted severely the output of a steam plant. Another example might be the availability of coke-oven gas, which nowadays in some regions is no longer used as town gas. This gas can be used advantageously as a gas-turbine fuel in a combined-cycle plant of the recuperation type.

Referring to the recuperation cycle, as defined by the author, I would suggest that for large utilities and for base-load operation this configuration is not very interesting. This is mainly because of the low unit output of the gas turbines. At the moment, we have gas turbines of 50 MW, which can be incorporated in a combined-cycle plant to give 75 MW, if a non-fired waste-heat boiler is used. In the near future, with gas turbines of 100 MW, it will be possible to have a total output of 150 MW in the combined-cycle plants of the recuperation type. However, for the larger utilities base-load power stations of such an output would not today be given serious consideration.

Within the restriction of base-load power plants, in my opinion, combined cycles are very interesting where extension of an already existing steam power plant is under discussion. Older steam plants work with a low efficiency. Adding gas turbines to such steam plants can result in a considerable gain in efficiency. Later, when the steam plant becomes completely obsolete, it can be scrapped and the gas turbine be transferred to peak-load coverage. This procedure can be exploited profitably by utilities of varying size and also by private industrial generating stations.

I was very interested to hear that once again the C.E.G.B. is looking at a combined cycle, in this instance for the Deptford station. I assume that because of the investment costs, this plant will have only one generator. That means that the two cycles, viz. gas turbine and steam turbine, cannot operate independently. This has some unfavourable impact on the availability of the plant. Concerning this proposal I should like to ask the author of Paper 76 three questions: Is the gas generator shown in Fig. 76.8 in fact representing a number of aero-derived gas generators? During the six months which elapsed between the submission of the paper to the Institution and this conference has the situation changed? Is the proposal still under serious consideration?

With reference to Paper 89 I would stress again that the plant at Hohe Wand is not representative in a general sense. From this paper we learn that a comparative study was made of a pure steam-turbine plant and two different combined cycles with and without preheating of the compressed air. These three possibilities are not strictly the only existing ones.

In the study the assumption was made that a restricted

amount of cooling water should be heated from 8°C to some 15 or 16°C, according to my calculations from Fig. 89.1. Such a small temperature difference guarantees a low condenser pressure and results in a good thermal efficiency of the cycle. My question is now: Why was an alternative proposal for a steam plant generating 75 MW with a higher but nevertheless tolerable temperature increase of the cooling water, giving a higher condenser pressure and a lower thermal efficiency, not examined? Such a plant would have been cheaper than the plant actually built.

Ten years ago when the Hohe Wand station was laid out, because of the low unit outputs of the gas turbines it was not feasible to examine the simple recuperation combined cycle with a non-fired waste-heat boiler. Such a configuration, with the same amount of cooling water, could give 40–50 MW of steam-turbine power output together with 80–100 MW in gas-turbine power output, resulting in a total station power output of 120–150 MW at an extremely low investment cost; but again with a somewhat lower thermal efficiency.

When the supply of cooling water is limited, now that the unit outputs of modern gas turbines are so much higher, such regeneration cycles should always be examined.

K. Gorter Arnhem

Paper 97 says (p. 119) 'Both fuel and air being gaseous, allowing simple flow measurement and proportioning together with the ease of combustion of gas as a fuel, enables us to obtain very low percentages of maximum load as a minimum load.' I should like to point out the danger of making air-flow measurements too simple.

Both 1 300 000 lb/h Benson boilers of our Flevo power station designed for oil and natural gas, but practically always gas-fired, are provided with safety devices that cause an automatic shutdown if air flow is less than 20 per cent or air/fuel ratio is less than 85 per cent of stochiometric conditions or gas pressure is too high or too low. Incorrect measurements of air flow could therefore either prevent a necessary shutdown in a dangerous fuel-rich situation or bring about unnecessary load-shedding of the units. A higher standard of measurement is required than in the days when the measurements were used only for automatic control which usually had in itself a possibility of correction.

The boilermakers frequently obtain flow measurement by taking the pressure drop over a venturi-like element placed in the air ducts where this is possible, and therefore often in a far from ideal position. These devices show rather irregular characteristics and they are very inaccurate at lower loads.

Sometimes the pressure drop over a certain part of the installation (e.g. air nozzles of the burners) is used as a flow signal. It is clear that in this case the part in question may not show any deformation during service. Moreover these measurements can be unreliable at lower loads.

For the regulation of the air flow one still meets air

registers of a relatively primitive construction and consequently of a gradually changing characteristic.

It is obvious that for the correct and safe firing of gas the whole chain—measuring device, automatic control system, control apparatus including measuring devices, and safety systems—should be designed integrally at a certain quality level. We all know that there is no point in connecting a refined electronic system to inadequate measuring apparatus. In my opinion it is advisable to pay attention to this matter at an early stage of boiler design. If this is done I can agree with the authors of Paper 97 that natural gas can be fired in a relatively easy and safe way.

I would conclude this part of my contribution with the following three remarks.

(1) With our outdoor boilers we had the problem that the transmitters and piping of air-flow and air-pressure measuring devices gradually filled up with condensate or with a sulphurous solution when oil was being fired. (Sulphur is picked up in small quantities in the Ljungström air-preheater.) The solution was to pass continually a very small quantity of pure dry air through the system.

(2) In the gas system, although our natural gas is technically dry, we found accumulation of moisture and corrosion in dead corners.

(3) It is useful to place a filter between the turbine gas-meter and the inlet valve of the boiler for we once found the blades of the turbine meter in our regulating valves!

Now I should like to ask a question about Paper 84.

On p. 94 the starting-up of a unit in one hour from light-up to full load after an overnight shutdown is mentioned.

As this is, in my opinion, a rather rapid pick-up of load and therefore very interesting, I should like to know some further details.

(1) I understand that, as the main and reheat steam temperatures are maintained only between 70 and 105 per cent load, the superheaters are situated in a position that is relatively favourable during firing up.

What are the criteria used to determine the rate of increasing the fuel input? Are they, as in our station, the gradients of superheater and reheater tube metal temperatures?

(2) How quickly after lighting-up are feedwater control and steam temperature controls switched to automatic?

(3) What is the maximum difference between turbine metal temperatures and steam temperature that is allowed at turbine run-up?

(4) How long does it take to run up the turbine to 3000 rev/min?

(5) Did the turbine makers use special constructions to meet the demand for short times for run-up and loading?

W. J. Hoornweg Arnhem

Commenting on Paper 72 on 'Commissioning of large power stations' I should like to refer in particular to the statements made in this paper on load rejection tests. I fully agree with the author that such tests are necessary,

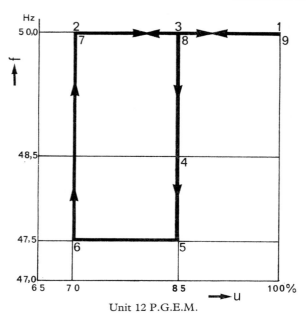

Unit 12 P.G.E.M.

Fig. D4. Conditions under which tests were carried out

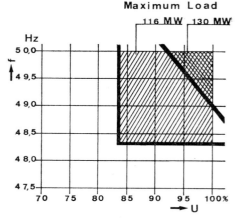

Gelderland 12 with two boiler feed pumps and one extraction pump.

Fig. D5. Operating limits

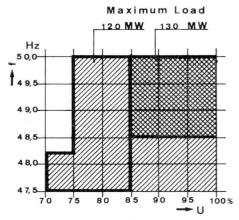

Gelderland 12 with three boiler feed pumps and two extraction pumps.

Fig. D6. Improved operating conditions with all auxiliaries running

but they should, in our opinion, be carried out on every new turbo-generator unit and should even be repeated, up to full load, after every major overhaul.

Another type of test which we consider of vital importance relates to the abnormal operating conditions to which the station auxiliaries may be subjected with regard to voltage and frequency. These conditions were defined in Paper 73 on electric power supply in the Netherlands. Tests to confirm reliable operation under these abnormal conditions were recently carried out on a 130 MW gas-fired unit at Nijmegen.

The conditions under which the tests were carried out are summarized in Fig. D4.

In this test programme we start at point 1 at normal conditions of 50 c and 100 per cent voltage. The load is 30 per cent, i.e. 10 MW. From here we planned to go to point 2 (70 per cent voltage) and then back to point 3. Here the conditions are brought to $48\frac{1}{2}$ c and 85 per cent voltage and a load of 130 MW to point 4. According to the defined conditions it is necessary to maintain full load at this point; furthermore the motors must be able to start again for a new unit start-up.

After this test we reduce the load again to 40 MW and go to points 5 and 6 ($47\frac{1}{2}$ c, 70 per cent voltage). At this point we bring the load to about 100 MW. Then we raise the frequency to see what is possible with normal frequencies and low voltage (point 7).

The next figure (D5) shows the operating limits, which were found as *results* during this test. As may be seen with the normal auxiliaries running (two 60 per cent boiler feed pumps and one 100 per cent extraction pump) full power is only achieved in a limited area, mainly because discharge from the condensate pump is reduced.

The last figure (D6) shows improved operating con-

ditions with all auxiliaries running (three boiler feed pumps, two condensate pumps). In this case full power is maintained with voltages dropping to 85 per cent and frequencies to $48\frac{1}{2}$ c. A reduced load of 120 MW was achieved with a frequency of $47\frac{1}{2}$ c and a voltage of 70 per cent in the hatched area.

Finally, we found that it is possible to start up all electric motors for the auxiliaries at the point of $48\frac{1}{2}$ c and 85 per cent voltage.

Answering a question of B. Wood I can add that the conditions under which the tests were carried out were as follows: the production unit was in this case coupled with the high-voltage grid as is normal, but the auxiliaries were fed by a separate small turbo-generator. Voltage and frequency were controlled by this turbine.

I turn now to Paper 97 and in a few words would say something about our experiences of measuring the volume of natural gas for our power stations at Lelystad and Nijmegen. It is clear that we must measure this volume firstly because we need to know how much the customer has to pay to the gas company and secondly so that the customer himself can justify the quantity in relation to the electric energy generated.

However, it is our experience that, with the apparatus at present available in the installation, the accuracy of measurement of volume is not more than about 2 per cent (that is to say plus or minus 2 per cent). You will understand that this is not sufficient accuracy for such large quantities of gas, say about 40 000 Nm³/h (standard cubic metre) for a 130 MW unit and about 55 000 Nm³/h for a unit of about 200 MW. The flow meters (turbine meters) are installed in the gas supply system by the gas company. The counting of this meter can be corrected by use of a pressure gauge and a temperature reading. Usually the customer has an orifice meter in the gasline to his boiler.

Our views about the deviation in the results of measurements are based upon:

(1) A comparison of the flow meters by charging the measuring and regulating station at a constant turbine rate. We found deviations of between 2 and 3 per cent.

(2) When the fuel is changed from heavy oil to natural gas, the heat consumption of the unit increases. According to our calculations it should decrease.

(3) It is not possible for the customer to check the flow meter (which is calibrated by the Dutch Calibrating Authorities). What we can do is to check the pressure gauge. The problem here is that the deviation of the manometer changes in course of time; we found for example 0·5 per cent in 6 months.

In my opinion it must be possible to reach a total accuracy of less than 1 per cent with a calibrated flow meter and an accurate pressure gauge.

I know that Gasunie is working on an improvement of the total flow meter arrangement; for example they have a test installation in the Diemen power station, in which measurements take place on the high-pressure side, and furthermore Gasunie is working on a development of density measurement instead of a pressure and temperature measurement.

We have every confidence they will be successful in achieving the improvements. We feel that the turbine meter as such is correct, but that it is possible that its place in the installation as well as the way it is used may cause the deviations. As far as I know, other Dutch power stations meet problems of the same kind. So I ask myself whether these problems exist also in other countries and whether perhaps somebody can tell us of the situation there.

R. M. Inches Gosport

I should like to comment on two points which have been made in Papers 76 and 89 and also in the discussion which has already taken place, concerning which points I believe the Royal Navy has some special experience.

The first point is the problem of noise. I believe it is an oversimplification to represent this in the form of a single dB value. The R.N. has learnt that the peak frequency, the width of the band over which the level is too high and even the nature of the noise, all play a part in determining both the seriousness of the problem and the best means of tackling it. The R.N. has also found when contemplating noise reduction measures that it is of value to consider the nature of the task or tasks which will have to be carried out under the influence of the noise. The performance of tasks involving any decision-making, for instance, must be expected to fall off under the influence of noise much more rapidly than the performance of routines or the obtaining and recording of information.

The second point is the effect of environmental conditions on gas-turbine performance. The theory behind this is of course basic and well known. What the R.N. has found, which might not be so well known, is that there is no mitigation of the theoretical effect by practical circumstances: the effect, in all its forms, comes fully up to expectations! In a beneficial sense, this may be making a contribution to the most commendable performance of the gas turbine reported in Paper 89; both the average ambient temperature and the air purity look to be favourable in that location. When comparing with this other existing plant elsewhere, or when contemplating the installation of plant of this kind in a very different environment, we should check the effect of this factor. I have in mind particularly schemes for setting up generating plant in the middle of cities—Edinburgh had been mentioned. While some might say that its climate is not very different from the climate at a reasonable altitude in the Austrian Alps, its nickname of 'Auld Reekie' suggests a considerable difference in air purity! However, this in turn is affected by the proportion of fuel burnt locally that is smokeless.

A. Kantner Essen

With reference to Paper 61 on air-pollution problems from large power stations I should like to draw attention to a problem we are very concerned with in Germany at present. It is mentioned in Paper 61 in one small paragraph. It is desulphurization either of the flue gas or of the fuel oil.

The results of attempts to remove sulphur oxides from the flue gases can be summed up as follows. The problems are mostly converted into those of disposal of some kind of material or of water pollution. Therefore for some processes the method of recycling with make-up of the materials used in a central plant is proposed. This again raises problems of transport, etc., but in this way it may be possible to produce highly concentrated sulphuric acid or even elementary sulphur.

In every case it is to be expected that the flue-gas treatment plant will be as large as the boiler itself. I think no power engineer likes to have a chemical plant like this in his station. If you calculate the production cost of the electric energy, you get a rise of about 0·4 Cent per kWh based on 4000 to 5000 h of operation.

This amount increases if the hours of operation go down, which is to be expected when nuclear power stations take the base load in the future.

Therefore I think it is worth while to pay more attention to the reduction of the sulphur content of fuel oils at refineries. This problem is urgent because the decision on the point at which the sulphur should be removed (whether from the oil or the flue gas) could be a final one. The reason is that the devices for the make-up of the recycled materials in the flue-gas treatment or for the reduction of the sulphur content in the fuel oil should be large enough to reduce costs. I think this problem should be dealt with on an international basis.

D. J. Moore Leatherhead

There are a few minor criticisms of Paper 61 that one could make. The construction of a wind-tunnel model and the accompanying experimental programme are costly. In 'normal' topographical situations, the differences between the actual surface concentration and that calculated from a simple model such as that described in equations (61.5) and (61.6) are likely to be no more than those due to variations in the ratio of the values of σ_y and σ_z with distance due to the fact that the atmospheric stability is seldom truly neutral [1].

I would therefore suggest that one should not normally use a wind-tunnel model, although in the case described the topography was sufficiently 'abnormal' to justify model tests.

A second point is that the degree of downwash observed in model experiments (Fig. 61.6) is usually much more severe than is observed in real life. This is due to three factors:

the Reynolds numbers of the flow round the stacks are very different;
the vortex behind the stack is more pronounced in the model because of the absence of long-period turbulent fluctuations in the tunnel;
the buoyancy effect may not be correctly scaled [2].

Downwash is therefore better studied in a separate experiment where a much larger stack can be used to examine the flow near the stack in detail, rather than in a topographical model where the stack scale is much reduced.

It would also be interesting to learn the amounts by which the plumes from the 175 m and 185 m chimneys at Wanganui dipped.

REFERENCES

(1) MOORE, D. J. 'The distribution of surface concentrations of sulphur dioxide emitted from tall chimneys', *Phil. Trans. R. Soc.* Series A 1969 **265**, 245.
(2) OVERKAMP, T. J. and HOULT, D. P. 'Precipitation in the wake of cooling towers', *Atmosph. Environment* 1971 **5**, 751.

G. Roozendaal Arnhem

I wish to comment on Papers 72 and 84.

After commissioning of the two sections of the Flevo power station 25 control test runs were made towards the end of 1969 and during the summer of 1970. This power station consists of two sections each of 193 MW, burning natural gas or heavy fuel oil. In accordance with what the authors of Papers 72 and 84 said about the growing importance of the control and automation equipment caused by rapidly rising labour costs, growing complexity and the changing role of the grid during the lifetime of the station, the requirements for the control equipment of these boilers were exactly formulated.

An engineering order was given to the boilermaker. The supplier of the equipment was chosen by the Provinciale Gelderse Elektriciteits Maatschappij in close co-operation with the boilermaker. The boilermaker was responsible for the good working of the boiler control equipment. As a result of these exactly formulated responsibilities and requirements, control test runs had to be made as a part of the commissioning as soon as the boiler controls were adjusted. The boilers are Benson boilers with a circulating pump for loads under 38 per cent of the maximum continuous rating. The steam pressure is 180 atm and the steam temperature is 535°C for both the live and the reheat steam. The turbo-generator has an electrical turbine control. The boiler controls are also electrical with electrical Ferraris type actuators.

As it is, of course, impossible in the five minutes permitted for these comments, to show you all of the results, I have chosen a sudden load fall, a sudden load rise, a quick load pick-up and a quick load fall with gas firing. Fig. D7 shows the steam pressure at the boiler outlet as a result of a 12 MW/s drop in load from 190 to 175 MW, followed by a decrease in load of 12 MW/min to 140 MW. The maximum pressure rise was 9 atm, not exceeding the limit of 10 atm, above which figure the high-pressure bypass system would have opened. The maximum live-steam temperature changes remained within a band of 4 degC and the reheat temperature within a band of 10 degC. The oxygen content of the flue gases changed from 0·6 to 0·4 per cent.

Fig. D8 shows the results of a rapid pick-up of the load. Here the pressure-drop limiting device of the turbine did

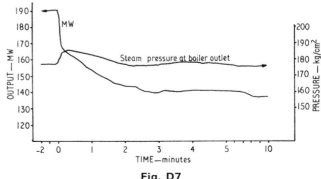

Fig. D7

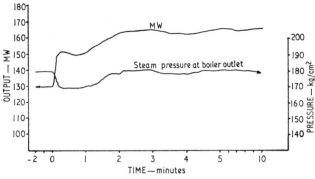

Fig. D8

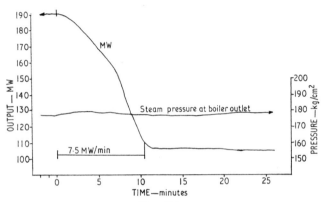

Fig. D9

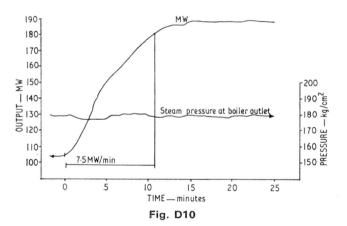

Fig. D10

not come into operation. Maximum steam-temperature variations were: +7 degC for the live steam and +12 degC for the reheat steam. The oxygen content changed from 0·5 to a maximum of 0·8 per cent.

Fig. D9 shows the results of a load change of 7·5 MW/min from 190 to 110 MW. Here the pressure remains within a band of 3 kg/cm². Temperature changes remain within a band of 5 degC for live steam and a band of 10 degC for reheat steam. The oxygen content changed in a band of 0·4 to 0·8 per cent.

Table D1

Change of output	Rate of change, MW/min	Rate of change, per cent/min
Between 40 and 75 MW .	4	2
Between 75 and 110 MW	6	3
Between 110 and 193 MW	12	6

Fig. D10 shows the results of the same load-change gradient of 7·5 MW/min from 105 to 180 MW. The steam temperature remained in a band of 3 kg/cm². Temperatures remained within a band of 5 degC for live steam and 10 degC for the reheat steam. The oxygen content changed from 0·5 to 1 per cent.

As a result of the test runs we were able to define a degree of flexibility of this plant, that is to say a maximum practical control speed, at which no limitations of the plant in pressure fluctuations, steam temperatures or oxygen content of the flue gases are exceeded.

The resulting gradients are given in Table D1.

The results for oil showed that the same flexibility could be achieved.

J. O. Stephens U.S.A.

In commenting on the 'Future prospects' section of Paper 76, I would say that there exists an important commercial reason for the slow acceptance of the combined cycle in the U.S.A. in addition to those reasons listed by the author. If we analyse the combined-cycle plant by individual equipment supplier and the effect upon the financial status of each supplier for every kilowatt installed, we get the figures given in Table D2.

In addition, the consulting engineer has a reduced incentive in that it is more difficult to sell his services to the utility for the same percentage payment when a portion of the plant is a prepackaged gas turbine with little outside engineering required.

The gas-turbine manufacturer is the only supplier who has a real interest in promoting combined cycles, so the slow acceptance in the U.S.A. is understandable. This has led both of the U.S. large electrical manufacturers to consider making boilers of the waste-heat or low-level firing type. In addition both (plus several other jet-type gas-turbine manufacturers) are now offering complete 'turn key' (completely installed) combined-cycle plants.

Table D2

Equipment manufacturer/supplier	Effect of large combined cycle on income for every kilowatt installed, per cent
Gas turbine	increase 15–20
Boiler	decrease 15–20
Steam turbine	decrease 15–20
Condenser	decrease 8–12
Generator (two generators) . .	increase 2–3
Switchgear (two sets of switchgear)	increase 2–3
Plant auxiliaries and building .	decrease 15–20

All of these factors plus the lack of long-time operating history for large gas turbines have contributed to the slow acceptance of large combined cycles. But I would add that the Westinghouse 301 gas turbine in the San Angelo plant of West Texas Utilities operated all of last year with a zero forced outage rate, or 100 per cent reliability.

A more recent activity of the combined cycle in the U.S.A. is pointing toward a 50–50 per cent steam and gas turbine because of the higher firing temperatures of the gas turbine which attains equivalent efficiency as the author of Paper 76 described plants. Furthermore, with regard to ecology, the 50–50 per cent type of combined cycle is one of the 'cleanest' plants that can be obtained.

B. van den Hoogen Leiden

In Paper 76 the author described the recuperative combined cycle on the one hand as a cycle with an attractive heat rate requiring relatively low capital investment costs, but on the other hand he noticed that although two combined plants in the U.S.A. are successfully in operation this has not led to repeat orders. I should like to add some remarks on these aspects and add the information given in Table D3.

As boiler manufacturers, especially in the field of waste-heat boilers, we are strong believers in the recuperative type of combined cycle. We base this belief on the fact that in addition to the feature that with these cycles attractive heat rates can be obtained these installations can be very

quick-starting too. A properly designed waste-heat boiler can easily meet the load variations and the starting procedure of an industrial gas turbine and this type of boiler can be started from cold in less than 15 min. A load variation of 10 000 kg/h of steam can be achieved in 6 s. These aspects might be of interest for mid-range applications for utilities and they proved to be very important to a large number of industries.

This brings me to my second remark, that Table 76.1 might give the impression that the recuperative type of cycle is an installation of rare occurrence. However, the total number of waste-heat boilers built or to be built to the design of the General Electric Company that produce steam to drive a steam turbine is 63.

Most of these installations have back-pressure steam turbines; among those 63 boilers there are seven units which are pure STAG cycles. I think therefore that the recuperative type of combined cycle has very good future prospects.

I should like to ask the author of Paper 76 what is his opinion of the recuperative cycle with regard to two aspects of which a considerable number of process industries have apparently realized the advantage already. The two aspects are (1) the ability to start quickly (2) the fact that a recuperative type of installation can easily be installed near the load centres, which is preferable to the building of big units that are half in the sea, which only results in an energy transportation problem.

Table D3. General Electric Company heat-recovery steam generators

Station	Gas turbines			Heat-recovery units					
	No.	Manu-facturer	Output, kW	No.	Type	Use	Steam, lb/in² degF	Rating, each, lb/h	Year of operation
City of Fairbanks, Alaska . .	1	GE	5 300	1	UF	St	610/750	38 500	1963
Community Public Service, Lordsburg, N.M. . . .	1	GE	11 500	1	UF	St	450/710	57 000	1964
					UF	FWH	FWH	22 700	
Borough of Lansdale, Pa . .	1	GE	11 500	1	UF	St	250/500	78 000	1964
Phillips Petroleum, Avon, Calif. .	1	GE	6 000	1	UF	Pr	275/450	57 600	1965
Std. Oil of Kentucky, Pascagoula, Miss.	2	GE	13 500	2	UF	St	600/650	81 900	1965
Union Carbide, Whiting, Indiana .	1	GE	13 500	1	UF	Pr	380/618	84 600	1965
					UF	FWH	FWH	29 400	
Union Carbide, Texas City, Texas .	1	GE	13 750	1	UF	St	610/750	77 700	1965
United Fuel Gas, Ceredo Station .	1	GE	10 500	1	UF	Re Stag	325/550	21 600	1965
U.S. Steel, Clairton, Pa . .	3	GE	14 500	3	UF	St	600/659	63 500	1965
Kellogg Co., Hastings, W. Va .	1	W	16 500	1	UF	OH	450/240	57·3 (10)⁶*	1966
Ottawa Water & Light, Ottawa, Kansas	1	GE	6 000	1	UF	Stag	422/800	47 600	1966
Southland Paper, Sheldon, Texas .	2	GE	13 500	2	UF	St	875/825	68 500	1966
Transcontinental Gas Pipeline, Tylertown, Miss. . . .	1	GE	8 000	1	UF	St	625/750	35 600	1966
Transcontinental Gas Pipeline, Billingsly, Ala	2	GE	8 000	2	UF	St	625/750	35 600	1966
Union Carbide, Texas City, Texas .	4	W	15 500	4	GF	St	625/750	160 000	1966
Union Carbide, Texas City, Texas .	4	W	20 000	4	GF	St	625/750	179 000	1966
U.S. Steel, Clairton, Pa . .	4	GE	19 000	4	UF	St	600/659	63 500	1966
Western Power & Gas, Great Bend, Kansas	1	GE	13 500	1	UF	FWH	FWH	470 000	1966
Wheatland Electric, Garden City, Kansas	1	GE	15 500	1	UF	St	400/650	70 500	1966
Wolverine Electric Corp., Inc. .	1	GE	23 000	2	UF	Stag	400/800	81 400	1966
					UF	Stag	102/339	13 159	

Table D3—contd

Station	Gas turbines			Heat-recovery units					
	No.	Manu-facturer	Output, kW	No.	Type	Use	Steam, lb/in² degF	Rating, each, lb/h	Year of operation
Allied Chemical, Baton Rouge, La .	4	GE	15 250	4	UF	St	850/750	83 700	1967
Kaiser Aluminum, Gramercy, La .	3	GE	15 250	3	GF	Pr	650/550	175 000	1967
Stauffer Chemical, Henderson, Nev. .	1	GE	15 250	1	GF	St	850/825	130 000	1967
Tennessee Gas, Gran Chenier, La .	1	GE	13 500	1	UF	St	405/750	46 100	1967
Transcontinental Gas Pipeline, Tylertown, Miss. . . .	2	GE	8 000	2	UF	St	625/750	35 600	1967
Union Carbide #11, Seadrift, Texas	1	GE	13 750	1	UF	Pr	640/SAT	92 500	1967
Carolina Power & Light, Moncure, N.C.	2	GE	17 300	1	UF	St	300/660	198 200	1968
Carolina Power & Light, Moncure, N.C.	2	GE	17 300	1	OF	St	300/660	280 800	1968
Duke Power, Mount Holly, N.C. .	4	WC	160 000	4	DFF	St	581/473	450 000	1968
Hess Oil, St Croix, Virgin Islands .	3	GE	15 250	3	UF	St, Pr	325/470	94 000	1968
Royal Dutch Salt, Netherlands .	1	GE(NP)	15 250	2	GF	Pr	427/572	178 900	1968
Southland Paper #3, Lufkin, Texas	1	GE	15 250	1	UF	St	600/750	84 100	1968
Tennessee Gas, Morehead, Ky .	2	GE	6 000	2	UF	St	475/770	40 800	1968
Humble Oil, Baytown, Texas. .	1	GE	17 500	1	GF, PE	St	625/750	234 000	1969
City of Clarksdale, Clarksdale, Miss.	1	GE	17 500	1	UF	Stag	450/808 109/334	85 600 14 300	1970
KZO, Amsterdam, Netherlands .	1	GE	17 500	1	GF	Pr	149/572	185 000	1970
Amercentrale, Netherlands . .	2	BB	20 000	2	UF FWH	St FWH	235/572 235/572		
South Carolina Electric & Gas, Parr, S.C.	2	GE	17 500	1	DFF, PS	St	425/750	367 000	1970
Texaco, Port Arthur, Texas . .	1	GE	17 500	1	GF	St	620/720	174 100	1970
Tri-Energy/TWA, Kansas City .	2	WC	17 800	2	UF	Pr	250/406	110 000	1970
Union Carbide, Ponce, Puerto Rico .	1	GE	13 750	1	UF	Pr	635/Sat	96 500	1970
United Fuel Gas, Ceredo, W. Va .	1	GE	10 500	1	UF	Re Stag	325/550	23 250	1970
Utah Power & Light, Salt Lake City, Utah	1	GE	17 000	1	GF, PE	Pr	160/423	225 000	1970
Worthington-Gulf Coast Aluminum, Lake Charles, La . .	3	WC	17 800	3	GF	St	865/830	165 000	1970
South Carolina Electric & Gas Parr #2, Parr, S.C. . . .	2	GE	17 500	1	DFF, PS	St	350/710	500 000	1971
City of Hutchinson, Hutchinson, Minnesota	1	GE	7 500	1	UF	Stag	407/798	49 000	1971
Royal Dutch Salt, Netherlands .	1	GE	17 500	1	GE	St	443/770	221 000	1971
Hess Oil, St Croix, Virgin Islands .	2	GE	17 500	2	UF	St, Pr	325/468	98 500	1971
Mines Dominiales P.A., France .	3	GE	17 500	3	GF	Pr	157/370	185 000	1971
Antar P.A., France . . .	1	GE	17 500	1	GF	St	725/815	88 300	1971
Société Chimique de Charbonnages, France	2	GE	20 150	2	GF	St	520/690	345 000	1971
Solvay et Cie, Belgium . .	1	W	27 100	1	F	Brine Heater/St	460/600	198 400	1971
River Works Utilities, Lynn, Mass	1	GE	21 700	1	RF/OF	St	650/850	190 000	1972
Kansas Power & Light Co., Topeka, Kansas . . .	2	GE	21 700	1	UF	Pr, St	400/700	184 400	1972

* Btu/h.

Type: UF Unfired	Use: St To drive steam turbine
GF Natural-gas fired	OH Oil heater (for heating process oil)
OF Distillate oil fired	FWH Feedwater heaters
RF Residual fired	Pr Process steam
DFF Dual-fuel fired (gas and distillate)	Stag GE combined cycle
PE Pre-evaporator	Re Stag Regenerative stag
PS Pre-superheater	

F. A. W. H. van Melick Geertruidenberg

Paper 76 gave us at a glance the specific characteristics and the different forms of combined cycles, together with some examples of running plants. I should like to add another example of application.

In the Amer Power Station of the Noordbrabantsche Electriciteits-Maatschappij, about 40 miles east of The Hague, we are about to commission a 400 MW unit. This unit has for its auxiliary service a gas turbine of 21 MVA with a waste-heat boiler in a recuperation cycle. In this way the main unit has an independent source for electrical power and steam for the auxiliaries.

Fig. D11 shows the Amer power station. The farthest building contains the units 1, 2 and 3 with a total output of about 435 MW. In the second building are the units 4 and 5, each of 230 MW, and the nearest building contains unit

Fig. D11. Amer power station

6, which will shortly be commissioned, as well as unit 7 which is still in course of erection. Units 6 and 7 each have a capacity of about 400 MW and each is provided with a combined cycle.

Fig. D12 shows the line-up of unit 6 in a horizontal section. The dotted lines represent the outlines of the gas turbine and waste-heat boiler, which are situated on ground level. The control and relay rooms are the level of the operating floor, just above the gas turbine.

Fig. D13 shows the vertical section of unit 6 and the situation of the gas turbine and waste-heat boiler.

Fig. D14 is a schematic diagram of the combined cycle.

The combination of a gas turbine and waste-heat boiler with its economizer, evaporator and superheater are well known in this composition. The waste-heat boiler has forced circulation and all tubes are of the fin type.

If the main boiler is fired with oil, the flue gases bypass the condensate heater and are mixed with the combustion air by the forced-draught fan. A control damper in the hot gas duct keeps the air temperature before the air heater at 80–90°C to prevent corrosion in this part of the main boiler.

If the main boiler is fired with gas, there is no need to preheat the combustion air, because the Dutch natural gas is sulphur-free.

The gases now pass the condensate heater in the waste-heat boiler and their temperature falls from about 200°C to 90–100°C.

The steam production of the waste-heat boiler is distributed over several uses. These are:

boiler feed pump;
oil heating and soot blowing (if boiler is oil-fired);
de-aerator;
preheating of condensate in the lowest high-pressure heater (A5) when there is superfluous steam.

The thermal and electrical diagrams have been composed in such a way that the combined cycle can be in service irrespective of whether the main unit is in or out of operation.

The electrical power will then be delivered to other auxiliaries of the power station or into the grid outside. If there is no need for steam but for electrical power, in that extreme case the boiler will be drained and heated up to

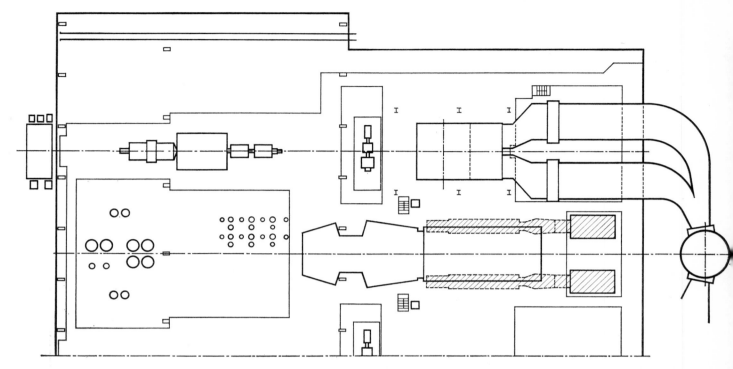

Fig. D12. Horizontal section of unit 6

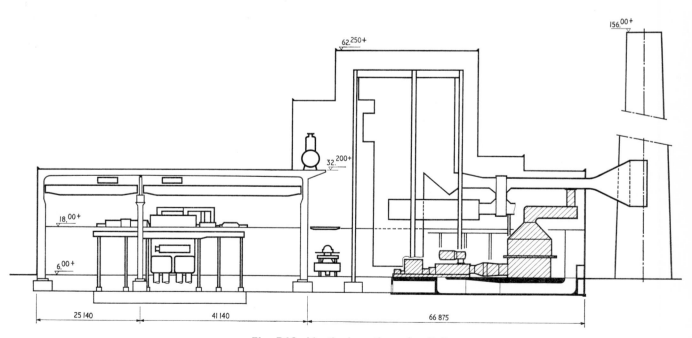

Fig. D13. Vertical section of unit 6

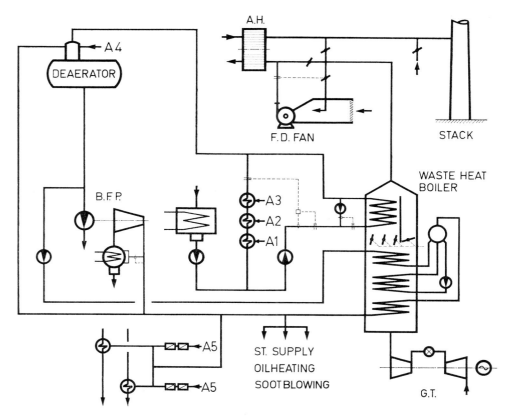

Fig. D14. Schematic diagram of combined cycle

360°C, the maximum temperature of the gases entering the waste-heat boiler.

Should the main unit miss the combined cycle, there will be a delivery of electrical power from outside, whereas the steam users are fed by bled steam. All bled-steam connections in this diagram are marked A1...A4.

Fig. D15 shows a graph of the steam production and consumption when the main unit is fired with oil.

To achieve the highest efficiency the gas turbine will always run at full load, independently of the load of the main unit.

At a higher load the steam consumption of the boiler feed pump will be completed by bled steam and at a lower load the superfluous steam of the waste-heat boiler will be consumed by the de-aerator feedwater heater and keep this vessel at low pressure.

The 10 kV bus-bars of the power station are connected by a cable feeder to the local grid outside, which is fed by the 150 kV grid.

In case of a failure in this grid, the power bus-bars and the gas-turbine generator would be implicated by this disturbance and at least a serious voltage drop would be the result.

To avoid any impact from outside on the bus-bars and grid for the auxiliaries of the power station, a short-circuit limiting coupling (SLC) has been installed. A short-circuit limiting coupling is a resonant link which

offers a low impedance and low loss to normal load current but reverts automatically and instantaneously to a high impedance state under fault conditions. In case of severe short-circuit faults in the 150 kV grid under the worst condition of full power station auxiliary load the voltage

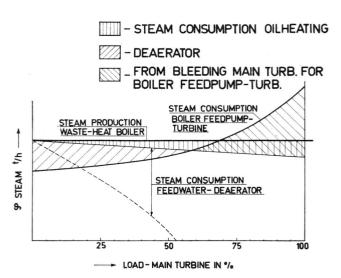

Main unit fired with oil.

Fig. D15. Graph of steam production and consumption

of the 10 kV bus-bars will be maintained at 75 per cent of normal voltage.

Besides the advantage of an independent electrical power and steam source for the auxiliary service of large units, this combined cycle makes it possible to heat and de-aerate the feedwater prior to the start-up of the main unit and fill up the main boiler with its own turbine-driven boiler feed pump. At the same time the flue gases from the waste-heat boiler can pass the main boiler furnace to warm up and are even sufficient to start firing and to keep the boiler at lower load without use of a forced-draught fan.

Instead of an auxiliary boiler the waste-heat boiler has already been in service for the same commissioning duties as are described on pp. 45 and 46 of Paper 72, and in our case with the same result in saving many weeks of commissioning time, because the heat for degreasing, chemical cleaning and passivation of both the waste-heat boiler and the main boiler was delivered by the gas turbine, whilst the steam from the waste-heat boiler was being used for driving the boiler turbine feed pump during the purging of the main boiler and steam piping.

In my opinion, the application of the combined cycle in this way improves the reliability of large units at no higher specific cost and at about the same heat rate.

F. Walker Glasgow

The experiences related in Paper 70 are all too familiar to power station commissioning engineers and operators. I should like to ask the following questions.

Has the use of television been considered for boiler water-level gauge glasses? It gives remote visual sighting and is inherently fail-safe.

Paper 71 dealt with the use of simulators for training operators. Is it not possible to apply this method to merchant navy ships, perhaps by the main shipping companies sharing a facility? It might also be useful to second engineering officers to modern oil-fired power stations to obtain experience of automatic and computer control, data loggers, etc.

There was a great deal of interesting and useful information in Paper 71 and it was unfortunate that the time available did not permit amplification of many points. The principle of specialist teams dealing with control and instrumentation is heartily endorsed and practice ashore is proving that more must be done in this direction.

The use of special commissioning teams is a good solution to many problems and is being applied to South of Scotland Electricity Board power stations. The main difficulty ashore is retaining the right type of engineer in a commissioning team for more than two years. The Royal Navy machinery trials unit must also suffer from this problem because officers are normally drafted to other appointments at intervals of about 2½ years.

I am envious of the elaborate simulators used to train M.T.U. and operating staff. In commissioning a power station the first unit away acts as a training simulator!

Would the author of Paper 71 please comment on this problem of maintaining an efficient team with constantly changing staff and also answer the following question? As R.N. boilers and turbines are now comparable with high-pressure power station plant what methods are adopted for chemical cleaning, steam purging and feed quality control during initial start-up?

I turn now to Paper 73 and especially to the section dealing with conditions of low frequency and low voltage and their effects on power station plant. In the South of Scotland Electricity Board we are very interested in these conditions and tests have been, and are being, carried out to determine the effects of low frequency and voltage on power station turbo-generators and auxiliary plant. The author will no doubt know that we attempt to maintain frequency at 50 Hz by manually shedding load in five blocks of approximately 20 per cent each and that in the event of a rapid frequency drop automatic relays operate to shed 20 per cent at 48·5 Hz and a further 30 per cent of the demand if the frequency reaches 48·0 Hz. We are required to maintain nominal voltage at the consumer's terminals within 6 per cent.

The S.S.E.B. are concerned that, in the extreme event of a complete system collapse, power stations should be able to retain at least one turbo-generator running at synchronous speed, supplying its own and other station auxiliaries in order to ensure a rapid return to power. In Paper 72 I mention a load rejection test on a 200 MW set which proved that the turbo-generator could supply its own auxiliaries through the unit transformer for at least 90 min. More recent tests have been carried out on a 300 MW machine to obtain information on the possibility of establishing one out of four 300 MW units at Cockenzie power station as a 'house set' in the event of emergency conditions such as an imminent or actual system collapse.

To date tests have proved that from time zero a 300 MW unit at full load can be clear of the bus-bars and carrying 12 MW load in under 60 s; the set then ran at 30 MW for 4 h without any difficulty. Tests are continuing in an attempt to decrease the time required to about 20 s.

Could the author of Paper 73 please comment on the following:

(1) Although the Netherlands power station plant is designed for the three extreme conditions given under A, B and C in the paper, would plant be operated under the conditions in A and B for the times given, and what effect would these drastic reductions have on consumers' motors and other equipment? Would not quick action be taken by load shedding to restore frequency and voltage to normal?

(2) In event of emergency conditions would the Netherlands transmission system remain fully interconnected, or would sections be isolated in an attempt to maintain normal frequency and voltage wherever possible?

(3) Could any details be given of tests carried out on power plant in the Netherlands under conditions of low frequency and voltage, and of tests which may have been done on establishing a 'house set' at a power station,

i.e. in emergency holding one turbo-generator at synchronous speed to supply station auxiliaries in readiness for a rapid return to power.

(4) The Central Electricity Generating Board install aircraft-type gas-turbine generators of about 60 MW at large power stations to supply station auxiliaries in emergency conditions of low frequency and voltage or system failure. Have the N.V.S.E.P. considered this possibility for Netherlands power stations?

The following comments and questions relate to Paper 84.

I note that the same problems with boiler control systems appear on each system commissioned, mainly spray control, drum level measurement and feed control. A new look at, and some detailed investigation into, these problems are overdue and some rethinking of logic is needed. Much of the trouble associated with control systems comes from the plant sensing device, particularly limit switches. The control system and plant designers might look at this jointly to achieve a satisfactory solution.

It is important that any auto-sequence control system, e.g. turbine automatic control equipment, should provide diagnostic information for the operator to inform him what is wrong if a sequence stops at any critical point. All too often the operator is faced with a sequence failure resulting in lengthy plant inspection before he knows what has gone wrong.

Are we tending to use automation unnecessarily? Two examples will show why I ask this question.

(1) During sequence starting of induced-draught fans which includes lubricating-oil pumps with feed-back of pump motor start, pressures, temperatures, cooler conditions, etc., there is ample time for an operator to start the oil pump and check all information at the fan site prior to the sequence starting of the fan motor, operating brake and dampers. Why include unnecessary equipment which increases start failure rate?

(2) Electronic governors on turbo-generators result in additional complication and increased fault risk. Is it right to change from the conventional type of governor in order to obtain the ability to choose settings when such a change increases the fault risk?

Finally, I wish to discuss Papers 76 and 89.

The South of Scotland Electricity Board have for some years operated three (4×17.5) 60 MW nominal aircraft-type gas-turbine units installed for peak-load lopping and emergency supplies. At present any interest in gas-turbine plant, either straight or combined cycle, would be only for peak loading and I would estimate requirements for peak-load units at about 1000 h/year with running between 12 to 17 h/day when required to cover peak periods which are now a flat-topped curve of long duration.

To consider gas turbines, with or without combined cycles, would require units of 70–100 MW output, at low capital cost with assured reliability and high availability, low maintenance costs and either fully automatic or requiring few operating staff. For peak operation efficiency and fuel costs are not too important, especially as maintenance and costs of operating staff may easily outweigh efficiency gains from more complicated combined-cycle plant. Another factor is the time to achieve full load from a cold start: anything over 20 min is liable to introduce problems of system control.

The availability of natural gas in Scotland could make gas-turbine combined-cycle plants attractive, especially if units of 300 or 500 MW were available, such units might be particularly valuable for two-shift operation.

The physical size and foundation requirements for gas-turbine units and combined-cycle plant are an advantage if one is considering the use of existing power station sites for replanting or as an addition, but as these sites are in built-up areas effective silencing would be required to ensure there were no noise problems.

O. Wiltshire Manchester

I would go further than the author of Paper 72 when he says that chemical cleaning should not be necessary for superheater and reheaters and would say that it is undesirable. It often results in the blockage of tubes by debris, particularly with hanging surface, and it seems to me that the better approach is to ensure that the tubes are clean when erected.

I am of the opinion that prolonged steam purging at high shifting factors for the purpose of obtaining clean target plates is totally unnecessary. The initial severe blows should remove any large debris and in my experience the target plates do not improve significantly after these initial blows. It should also be noted that the success of a blowing process can be in doubt because target plates can be indented by relatively soft debris, such as the magnetite formed during the blowing period, and even by water drops. Similarly it is essential to clean chemically the temporary pipework, or debris from this source can also continue to indent plates when the boiler itself is essentially clean.

In the U.S.A. chemical cleaning is often delayed until after 12 months of commercial operation, apparently without long-term ill effect. This could well save very valuable time in the early life of a unit and should on this score alone be considered in Europe.

I believe that if the author would look more closely at full-flow condensate polishing rather than shunt polishing he would be able to justify its installation on drum-type boilers with outputs as low as 200 MW. On a 120 MW unit, for which my firm are the consulting engineers, full-flow condensate polishing enabled correct water quality for full-pressure operation to be achieved in a few days without water dumping and this was coincident with a condenser leak. The polishing unit installed on this installation was a Permutit mixed-bed unit with a cation layer without a filter and the fact that it could clean up the boiler circuit while there was a condenser leak justified the higher cost as compared with a powdered-resin installation with a lower ion exchange capacity.

It would be useful if the authors of Paper 97 would provide more general information and background on the safety precautions adopted in Holland. More specifically I am interested to learn if these regulations are based on the rules of the National Fire Protection Association of the U.S.A. or if they have been evolved mainly from Dutch experience. Are the regulations enforceable by law?

I notice that on Fig. 97.4 a bleed valve to atmosphere with a flow indicator is provided between the double shut-off valves in the gas supply to the burners. I presume that the purpose of the bleed valve is to ensure that gas piping to the burner is clear of air before initial light-up. Is it usual in the Netherlands to time the bleed period or to rely on operating-staff experience to ensure that gas piping is correctly vented? In other parts of the world it is not unusual to establish a pilot flame at the end of the bleed and to interlock this flame with the master shut-off valve. Is this done in the Netherlands?

Is it Dutch practice to monitor each individual gas burner and igniter? There are other approaches to flame monitoring such as monitoring the igniters only, which are kept alight continually. It is argued by some that, if the igniter gas supply is independent of the main supply or supplemented, this is the safest arrangement.

Have the Dutch found it necessary to do more than relate the furnace purge to a number of air changes? There are suggestions that in the U.S.A. there is a move towards doing more, such as measuring combustibles in the furnace atmosphere.

Finally, what is the record to date with respect to furnace and associated explosions on gas-fired units in Holland? Should the authors have any first-hand experience of such events, a description of any conclusions reached as to the cause and the precautions necessary to prevent a recurrence would be a valuable addition to the paper.

B. Wood Esher, Surrey

My contribution refers to Paper 61. A great deal of model testing is of course now carried out because it is fashionable and clients like it. The last thing to be investigated is whether the test is valid. Validity is ensured only by dynamic similarity which calls for both Reynolds number and Froude number to be right, and this is usually impossible. Moreover, hardly anybody comes back afterwards and carries out checks on what actually happens when the power station is built, and this is the only valid check.

The attribution of downwash to the wide chimney in Fig. 61.3 is absurd. It is well known that the flattening of a smoke plume is largely a matter of momentum and since there is nearly always a 20 mile/h wind in England at 300 ft high (higher in New Zealand), the plume will in fact be practically flattened at any likely efflux velocity. There is no aerodynamic justification for arguing that chimney diameter has any bearing and the American idea of nozzling the tips of chimneys to give extra notional height by high efflux velocity has from the beginning been seen to be absurd except in still conditions when it is not necessary anyhow.

On p. 12 the author refers to the water-vapour discharge from wet cooling towers. He quotes a quantity of more than 3000 t/h. In fact the quantity is somewhat less, namely 2600 t/h. However, nearly all of this, probably 99 per cent or more, is in the form of invisible (true) water vapour and what one sees in the plume is the remaining 1 per cent, possibly 26 t/h of fine water particles. It is not clear what the author considers the relevance of the cumulus cloud to be. This could only form in windless conditions, which are extremely rare in England, and I have never seen it. No cooling tower has ever been damaged by lightning, its own or foreign. What we can definitely say is that the mass effect of 2000 MW of cooling towers close together will result in sufficient levitation from presence of water vapour and warmed air to punch through quite considerable barriers in the form of inversions; curiously enough in *The Times* of 20th April 1971 this is now quoted as a discovery by the General Electric Company of America. In other words large cooling towers can be credited with ventilating the surroundings.

With reference to the author's second point that it would be better if chimneys were on high ground, there is of course something in this but the suggestion that one would then resort to dry cooling towers is a *non sequitur*. We require for pumping the whole of the circulating water to a height of 30 ft less than $\frac{1}{2}$ per cent of the power generated. For the same loss of power we could pump the 1·1 per cent of cooling water necessary for make up to a height of 2700 ft! This shows that it is not the pumping of make-up water which prevents power stations from being built on hill tops, nor indeed far from water sources. It is probably simply the awkwardness of getting a way-leave. The cost of resorting to dry cooling on the other hand would be a sacrifice of about another 5 per cent relative to wet towers. The fact that a number of these have been installed in various places, including Rugeley, for purely experimental purposes (not with any economic justification) does not mean that they have any economic acceptability in countries like England where we are usually up to our ankles in water. In fact I have been trying to find economic justification for putting in dry cooling towers for the last 40 years but without success. The true dry condenser of the GEA (Luftkühler Gesellschaft), which is cheaper and easier to justify, has been available since before the 1939 war.

Authors' Replies

G. Nonhebel

B. Wood casts doubt on the validity of wind-tunnel model tests and states categorically that both the Reynolds number and the Froude number must be right. Opinion has changed. A recent paper (1) states 'all available experimental evidence suggests that Reynolds number is not a significant parameter in this case'. This was also the opinion of the research team in charge of the Inverkip model tests. Nevertheless J. F. de Lisle of the New Zealand Meteorological Service has commented in a private communication 'Unless one can model the planetary boundary layer in a more realistic fashion than was done in the New Plymouth tests (at the N.P.L.) I think the results are suspect. This can, of course, be done now in tunnels of very long section.' It is because of such doubts that the author advocates laying of smoke trails by aircraft over candidate sites during adverse weather conditions likely to accentuate pollution, e.g. high inversions. These trails should show up points not made clear during wind-tunnel observations. The S.S.E.B. staff concerned with Inverkip consider that the observations in the Bristol wind-tunnel have given confidence in the decision made on chimney height.

Only a few landscape wind-tunnel observations have so far been made and the power stations concerned will not be operating at full output until the end of this decade. The author suggests that the Institution arrange an international discussion thereafter on comparisons between observation and design studies. Three years' notice will be required for the collection of significant information.

The author cannot agree with B. Wood's comments on downwash. There are plenty of photographic records in the N.P.L. of the larger-scale models mentioned by D. J. Moore and staining of the top outside walls of chimneys can frequently be observed. The wide tower containing the Bankside (London) power station chimneys frequently exhibits downwash.

The amount of condensed water mist in the saturated plume leaving a cooling tower depends on the relative humidity and temperature of the ambient air and can be considerable where the humidity is high as occurs in many places in the world. The paper refers to visual pollution by long plumes of water mist. In his last paragraph B. Wood refuses to admit the growing shortage of water in land situations in almost every country, including England. Given the demand, inventive engineers will find ways of reducing the cost of air coolers for power stations as they have already done for oil refineries.

In reply to D. J. Moore, it should be noted that most new power stations will in future have to be built in difficult locations. No further information is available on the plume dip at Wanganui. The New Plymouth site has been chosen and the boilers will now be fired with oil or by the recently proved supply of local natural gas, and not with coal.

A. Kantner's valuable comment on the excessive capital charges of removing sulphur from flue gas, arising from progressive decreases in load factor after a few years, reinforces the author's view that desulphurization of oil by hydrogen is the right solution for oil-fired stations. Oil-treatment plants are operated continuously at full load and the resultant pure sulphur can be stored cheaply and is readily transported to distant markets.

It must again be emphasized that many laboratory and pilot-plant studies of methods for removal of sulphur dioxide from flue gases omit to take into account the trace impurities in the gases, such as chlorides, oxides of nitrogen, dust and soot. These can affect the continuity of the process and render products such as sulphuric acid and ammonium sulphate unsaleable unless subjected to expensive purification. For this reason the most desirable process for flue gas is also that which produces elementary sulphur through a series of intermediate steps, the last of which gives hydrogen sulphide by hydrogen reduction of absorbed sulphur dioxide.

The overall requirement of hydrogen is given by the equation

$$SO_2 + 3H_2 \rightarrow H_2S + 2H_2O$$

For oil hydro-desulphurization, the overall theoretical amount of hydrogen is one-third less, as shown by the equation

$$\begin{matrix} ---C \\ ---C \end{matrix} \Big> S + 2H_2 \rightarrow \begin{matrix} ---C-H \\ ---C-H \end{matrix} + H_2S$$

The answer to A. Kantner is clearly to encourage the oil companies to reduce the sulphur content of residual oil to perhaps 1 per cent. Demand for such large quantities of purified heavy oil will accelerate research on effective catalysts for hydro-desulphurization.

REFERENCE

(1) LORD, G. S. and LEUTHESSER, H. J. 'Wind-tunnel modelling of stack gas discharge', *Man and his environment* 1970 1, 115 (Pergamon Press, Oxford).

The following corrections to the paper should be noted.

Page (column)

1 (2) First full paragraph, last line: for 'at' read '='.
3 (1) Second full paragraph, second line: for 'east' read 'west'.
8 (1) Second line: for 'MWh)' read 'MW$_H$)'.
12 (2) Penultimate line: for 'dispersed' read 'dispersion'.
13 (2) Fourth full paragraph, line 5: for 'sulpher' read 'sulphur'.
13 (2) Reference (3) should be 1971, **5**, 155.

R. L. Dennett

As other papers presented at this Convention have indeed shown, many problems are shared by both marine and land-based engineers and operators. Manufacturing companies operating in both fields have the decided advantage of being able to draw on experience from either side. In this respect a recent appraisal of the use of television for sighting water level in boiler gauge glasses indicated that at the present time the disadvantages appear to outweigh the advantages in applying this to marine boilers where 'remote' locations are considerably closer to the machinery control station than they are in power stations. Direct vision can usually be arranged if given due priority.

The use of simulators for training operators, as shown by the author of Paper 71, can be extremely beneficial but until standard ships become the rule rather than the exception in the Merchant Marine the permutations of equipment used at present would make this difficult to establish by operating companies. Control manufacturers organize engineering courses and provide simulated control lay-outs of systems, thereby going part of the way to achieve this end. These courses can be augmented by visits to sites to see the equipment in operation.

In the foreseeable future it would appear that experience in commissioning a power station will also be true of ships: the first ship of a series will be the simulator for the second, the second for the third, and so on.

The following corrections to the paper should be noted.

Page (column)

27 (1) Line 12, should read: establishing the cause of and then eliminating defects
28 (2) Second paragraph of sections on 'Design', last line: 'providing installations' should read: 'providing automatically controlled installations'.
30 (2) Fig. 70.2. 'B.F. heater' should read 'O.F. (oil-fuel) heater'.

R. M. Inches

The problem which F. Walker mentions is a real one, and it is more correct to say that it has been minimized than to claim that it has been solved. To the achievement of this situation, several special measures have contributed.

Firstly, there is very close liaison between the appointing and drafting authorities on the one hand and the Officer-in-Charge of the machinery trials unit on the other, to provide the officer with maximum notice of impending changes and where possible allow him some say in the choice of replacements.

Secondly, though each of the three teams tends to specialize in certain classes of ship they all have an across-the-board capability. Using his knowledge of impending changes, the O.I.C. M.T.U. can thus make some anticipatory changes to minimize drops in effectiveness.

Thirdly, each team has a civilian element in it which gives continuity in general terms.

Fourthly, except in emergencies new arrivals to the teams either get a longer-than-usual turnover period or special training. (One must admit, however, that emergencies are relatively frequent!)

Finally, there is a comprehensive system of records which has been refined from experience. Together with rigorous insistence that all instrumentation required for trials information must be in working order before trials start, this provides even newcomers with a good basis for decisions.

Before answering F. Walker's second query, two points need to be made. The first is, that the author's present situation (serving in a non-technical appointment at CENTO headquarters in Ankara, Turkey) has him badly placed to answer this kind of question in detail. The second is, that not everyone would agree that the R.N.'s most advanced steam conditions (850 lb/in^2, 950°F) justify comparisons with current shoreside practice.

That having been said, the following general answer may still be of some interest.

Guidance on standards to be achieved, in all fields of engineering, is given to shipbuilders as part of the documentation which accompanies a contract. Decisions on procedures to be used to meet those standards are their responsibility. However, the knowledge that the M.T.U. will be involved in the acceptance process, and awareness of its very extensive experience, usually results in a fairly standard set of procedures being worked out in consultation. Chemical cleaning of boilers at some stage is most likely, but with pipe systems the emphasis is on cleaning before, rather than after, assembly. Special strainers may be fitted during the early stages of steaming. The importance of admitting only high-quality feed to a chemically cleaned boiler is fully appreciated and the products of the ship's distilling plant will only be used for this, normally, after the plant has satisfactorily completed its trials. On top of this, a thorough internal inspection of all boilers is carried out by the M.T.U. when it would be reasonable to expect them to have been properly 'conditioned'; if there is then any cause for concern, remedial steps are at once taken.

Should F. Walker wish to pursue this matter further, he should contact the Director of Engineering (Ships) in the Ship Department of the Ministry of Defence at Bath.

F. Walker

The author thanks W. J. Hoornweg for his valuable contribution and summarizes the results of some recent tests carried out at the South of Scotland Electricity Board's Portobello power station which may be of interest.

Portobello power station has three 60 MW unitized cross-compound turbo-generators. No. 1 turbo-generator was selected to carry a steady 60 MW load whilst No. 2 turbo-generator was used as a variable frequency and voltage supply source for both its own and No. 1 turbo-generator auxiliaries. Several tests were carried out but the ones worth mentioning are as follows.

Nominal voltage and low frequency. The frequency of No. 2 turbo-generator was reduced in 0·5 Hz steps from 50 to 47·5 Hz and at each step readings of load, frequency, auxiliary motor currents, generator winding temperatures, etc. were taken. The low frequency of 47·5 Hz was held for 10 min. Full load on No. 1 turbo-generator was carried without difficulty and no adverse effects were noted on any plant.

Nominal frequency and low voltage. The voltage of No. 2 turbo-generator was reduced to 90 per cent nominal volts, plant parameters being measured continuously. This 10 per cent voltage reduction caused difficulty on the unit transformer and many auxiliary motors which rapidly approached overload trip settings causing the tests to be concluded.

Low frequency and low voltage. On No. 2 turbo-generator the frequency was reduced to 49 Hz and then the voltage reduced to 90 per cent nominal. Frequency was further reduced in 0·5 Hz stages to 47·5 Hz and voltage then reduced to 85 per cent nominal. No difficulty was experienced in carrying full load on No. 1 turbo-generator and conditions on auxiliaries were satisfactory although the boiler automatic-control feed bypass had to be opened to maintain feed level on No. 1 boiler.

Comments on the results. The test results showed a close similarity to tests carried out by the Central Electricity Generating Board on a 120 MW unit some time ago. This would indicate that further tests on small and medium-sized power plants are not necessary but tests are being considered on large units where design margins may be less than those on smaller units.

G. Roozendaal's contribution was of particular interest because the S.S.E.B. have recently carried out response tests on 200 MW (Kincardine power station) and 300 MW (Cockenzie power station) units. Both stations have high-pressure plant with drum-type boilers fired with pulverized coal. Details of the Cockenzie plant and a commissioning summary are given in Paper 72. The tests were undertaken principally to determine the 'spinning spare' response of the units and also to determine the overall plant control response.

The tests showed that for rapid loading the units would give an initial increase of output of 8–10 per cent MCR in 10 seconds, the curves being fairly similar to those in Fig. D8. Further tests are to be carried out at Cockenzie

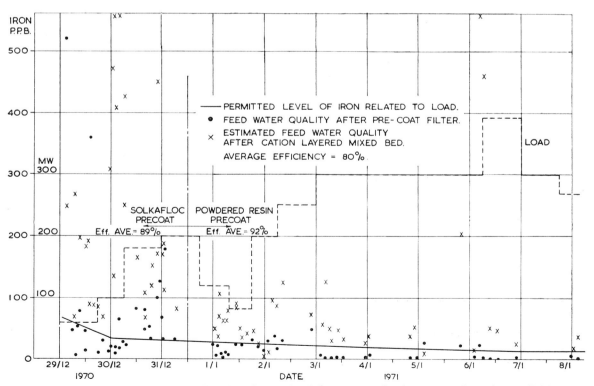

Fig. D16. Longannet power station, No. 2 unit. Initial start-up: feedwater quality after polishing

and also at Longannet power station once the plant is fully commissioned.

I was very interested in O. Wiltshire's useful contribution which gives me an opportunity to amplify relevant points in Paper 72.

I agree with his remarks that chemical cleaning is undesirable for superheater and reheater tubes. It is now the policy of the S.S.E.B. to ensure that the tubes conform to agreed clean standards when erected and no chemical cleaning is carried out after erection. It is worth mentioning that a technique has been developed in Belgium for high-pressure boilers which eliminates chemical cleaning completely. All boiler tubes are manufactured to a high standard of cleanliness, protected by grease, and the boiler is finally degreased when erection is completed.

With regard to steam purging, a method is used by the S.S.E.B. which successfully clears all pipework and produces clean etching plates after about 30 steam blows and takes only 2–3 days of actual steam purging. The method is to obtain the highest possible shifting factor for steam blows without target plates (ideally 1·6–2·0) and, when measurements show these have been achieved, target plates are fitted and steam blows continued at lower shifting factors, always greater than unity and normally about 1·3. Experience on the first three boilers at Longannet power station showed that acceptably clean etching plates were obtained after 2 or 3 blows. It is reasoned that if the pipework is proved clean at a shifting factor greater than unity it should certainly be satisfactory under normal working conditions.

I agree that full-flow condensate polishing should be installed on all modern high-pressure plant, drum or once-through boilers. The cost is fully justified and full-flow condensate polishing is essential if plant is to be on two-shift operation at a later stage in its life. I assume the 120 MW unit referred to is not high-pressure (2450 lb/in², 1050°F) plant because I doubt if the limits given in Table 72.3 could be achieved without a filter unit. Fig. D16 illustrates experience at Longannet with polishing plant. The solid line is the permitted level of iron related to load and it will be noted that whilst the pre-coat filter unit output meets requirements the estimated quality (based on previous measurements) from a cation-layered mixed-bed unit does not. Note that in the early stages of load raising, Solkafloc was used in the unit which, though not as effective as powdered resin, was much cheaper.

The following corrections to the paper should be noted.

Page (column)
40 (1)   First paragraph, last line: after 'drum' insert 'end'.

44 (2)   Second complete paragraph, last line: for 'boiler' read 'oil'.

46 (1)   Fig. 72.8. The dotted line marked: 'Condensate polishing plant outlet' should be at 0·01 p.p.m. as in revised figure below.

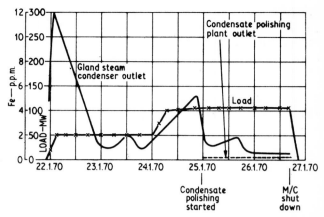

Fig. 72.8. Condensate polishing plant outlet during initial start-up of No. 1 unit at Longannet power station

49 (2)   Acknowledgements: After 'English Electric Company', add 'Foster Wheeler John Brown Land Boilers Ltd'.

G. A. L. van Hoek

The following are my replies to the questions of F. Walker.

(1) As stated in Paper 73, conditions (a), (b) and (c) on p. 60 are extreme conditions defined in order to make sure that production units, including their auxiliaries, will not cause a cascading failure consequent to a severe initial disturbance. Of course S.E.P. does not want to prolong a situation with abnormal frequency and voltage conditions, (a) and (b), unnecessarily. Automatic load shedding takes place, dispersed over the country, at 48·5 Hz (10 per cent) and 48·0 Hz (additional 20 per cent). To avoid unnecessary load shedding, production units should not worsen conditions at frequencies between 50 and 48·5 Hz arising as a consequence of an initial disturbance. In these circumstances production units should maintain their rated output.

If the frequency falls below 48·5 Hz automatic load shedding takes place all over the country. If there are gas turbines available, these will be started and national control will ask for further load shedding in the area where the initial disturbance took place. This additional load shedding is done manually. These measures take time, but usually less than the time delays stated under (a) and (b). The latter are maximum delays, just to ensure a sufficient safety margin.

(2) In the event of emergency conditions the Netherlands transmission system should remain fully interconnected; no deliberate 'islanding' takes place.

(3) As the requirements (a), (b) and (c) are applied to all new generators Samenwerkende Electriciteits-Productiebedrijven tries to keep all generators, each with its own auxiliaries, ready for a rapid return to the system rather than establish a 'house set'.

Details of tests carried out on power plant in the Netherlands under conditions of low frequency and voltage have been given by W. J. Hoornweg, commenting on Paper 72 in his communication, to which we refer here.

(4) Some Dutch power stations will have gas-turbine generators to supply station auxiliaries in emergency conditions, which can be a solution to (*a*) and (*b*) if frequency falls slowly, but not if frequency falls quickly.

Of course, starting up after a total system collapse can be ensured by such gas-turbine generators. However, S.E.P. tends more to the solution, mentioned under (1), of avoiding a total system collapse.

The following corrections to the paper should be noted.

Page (column)

53 (1) Fig. 73.4, vertical axis: for 'load' read 'consumption'.

56 (2) Second complete paragraph, lines 7 and 8: there should be no parentheses.

56 (2) Eighth line from the bottom: for 'efficient' read 'reliable'.

58 (1) Third complete paragraph, second line from the bottom: for 'efficiently' read 'sufficiently'.

58 (2) First complete paragraph, ninth line: for 'unit' read 'units'.

59 (1) First complete paragraph, line 6: for 'compulsory' read 'forced'.

M. C. van Veen

The following corrections to Paper 75 should be noted.

Page (column)

64 (2) Table 75.1. Third line from the bottom: for '0·05–0·90' read '0·50–0·90'.

65 (1) Tenth line from the bottom: for '1 MW' read '1 MeV'.

69 (1, 2) Fig. 75.10: for '8-MV' read '8-MeV'.

B. Wood

I am rewarded by the interest shown in my paper and grateful for comments and corrections from various people. Even though I had been adding to the list of combined plant in the interval between writing the paper and its presentation, it is clear that I had missed many and in particular grossly understated the number of recuperation plants in the U.S.A.

B. van den Hoogen lists no less than 37 recuperation plants, some with as many as four units, mainly with General Electric gas turbines. He also lists a large number where steam is raised for process, a practice which I thoroughly commend though had to exclude from consideration. He mentions that recuperation plant can utilize established sites near the load. This I agree is its major advantage since the saving on transmission about balances the lower cost per kW of the large remote plants.

F. A. W. H. van Melick mentions an interesting case at Amer power station of the recuperation cycle applied to a steam-turbine drive of the boiler feed pump (with additional steam use). This seems particularly valuable because the boiler feed pump can be started independently of the main 400 MW unit and also can run at variable speed (even though the gas turbine is synchronized). The waste-heat boiler I note serves as a donkey boiler and thus facilitates commissioning.

F. Walker mentions a need for units of 70–100 MW. This is actually a size readily available in the recuperation combination since a 50 MW gas turbine will make available about 25 MW from steam without supplementary firing and conveniently about 50 MW with. The combined efficiency is about 50 per cent better than that of the straight gas turbine at 75 MW and still appreciably better at 100 MW while the capital cost per kW on average rating is likely to be about the same as a gas turbine alone. There is no reason to believe that attendance is any problem since an unfired boiler needs substantially no attention. Starting time is for industrial gas turbines somewhat longer than for aircraft units but 20 min is an easy target. The gas turbine will indeed come on load probably within 10 min and the steam turbine perhaps 5 min later, quick starting and rapid loading being acceptable by reason of the low steam conditions. Such plant is economical for 1000 h duty per annum and upward at British prices for refined fuels, whereas the straight gas turbine (as installed in the U.K.) is probably not really justifiable for more than 500 h per annum average. There is no need to suppose that noise or pollution presents any new problem since industrial gas turbines are much less noisy than aircraft types which, as Mr Walker mentions, in ratings of 56 and 75 MW have already been installed in various parts of England and Scotland.

R. M. Inches also raises the question of noise and fouling by dirty atmosphere. In both respects it is necessary to differentiate between aircraft and industrial gas turbines. The noise level of the unprotected air intake is so high in either type that no one can be permitted in the air-intake house when a machine is running (there are other good reasons against this). The intake splitters have proved, however, in land installations quite effective in reducing external noise to an acceptable level. The exhaust noise is not likely to be troublesome where an exhaust boiler is fitted. The engine room with the aircraft type is not intended to be occupied. Control is exercised either remotely or from an adjacent sound-insulated control room. Sound-absorbent hoods can attenuate engine noise to a tolerable level for personnel wearing ear-guards. The noise is contained by the building in which windows are commonly absent, though examples are known of buildings with glass brick windows affording natural light and it is not possible to detect from outside that the plant is running. Industrial gas turbines produce 'white' noise, much as do steam turbines. Hence the sound level in dB is a reasonable measure and octave analysis is not necessary. Industrial gas turbines of large size are reasonably tolerant of air pollution. They are already widely used in cities, e.g. Bremen and Munich. Over 30 units of 50 MW and upwards will soon be at work in Europe so a gas turbine is no longer to be thought of as an unknown creature still in the trial stage. Total world installations of gas turbines is in excess of 70 million kW. About 85 per cent of these are of the industrial type.

M. Gasparovic's arguments have been partly answered by others. His doubts as to the cost of high-efficiency plants I can best answer by referring him to a paper by Cochet et al. (2) on the combined cycle unit for the Emile Huchet station of the French Mines Department. This new extension will contain a nominal 250 MW steam turbine which yields 294 MW with reduced bleed. There are three 16·5 MW gas turbines in parallel as well as a one-third duty forced-draught fan. This possibly somewhat costly solution results in high efficiency at part load and also high reliability. It is possible to operate with one gas turbine out of service. The total output at 343 MW is raised 37 per cent above the basic 250 MW while the investment cost is raised by only 20 per cent. The cost per kW of the increment is thus 12 per cent below the basic cost of the station. The reduction for scale would be only about 6 per cent on the two-thirds power law but would also imply departing from standard ratings of steam turbines and boilers.

A further very interesting point is that the reason for adopting combined plant at Emile Huchet is that coke-oven gas from nearby Carling is no longer acceptable to the gas authority which is now handling natural gas. The same situation has arisen in Germany; Siersdorf is being installed as a recuperation plant to fulfil the same function. Thus the combined cycle can be used in either form to put to efficient use waste gas which has suddenly become unacceptable to the gas undertakings. The gas turbine can make better use of a clean fuel than merely burning it under boilers and converts the energy into the saleable form of electricity. This is in the national interest but of course demands collaboration between monopoly undertakings. My first gas turbine project 33 years ago was exactly the same.

K. Goebel said he knew of nine high-efficiency combined cycle units in Germany but did not give details. It appears that the additional three 400 MW combined units at Gersteinwerk now announced will bring my six up to his nine, Emile Huchet in France being additional.

P. Cosar (whose contribution has not been recorded) mentioned that the quoted efficiency for Vitry, which I had taken as 39·6 on the lower calorific value, actually related to the higher calorific value. A correction has been made which brings the figure up to 42 per cent. He said also that the efficiency of Hohe Wand, quoted as 43·7 per cent, is strictly at the generator terminals and does not allow for auxiliary power consumption which might be 3 per cent. Hence I was wrong in stating that Vitry is for a comparatively low efficiency. In fact it is about as good as the best yet attained.

I appreciate the thoughtful contribution of J. O. Stephens (given orally at the meeting by W. H. Stinson)

explaining why the high-efficiency cycle is not making headway in the U.S.A.

REFERENCE

(2) COCHET, E., BILLARD, J., MIRIGAY, R., GOALVOUEDEN, J., LUGAND, P. and MOINAUD, C. 'Alimentation combiné au gaz de cokerie et aux bas produits houillers d'une installation productrice d'énergie integrée de grande puissance (Centrale Emile Huchet)', Paper 2.3–83 presented to World Energy Conference at Bucharest, 1971.

The following corrections to the paper should be noted.

Page 76.   Additions to Table 76.1

Installation	Year in service	Fuel	Gas turbine, MW	Steam turbine, MW	Efficiency on L.C.V.
Type 1: High efficiency Robert Frank, Germany (Preussag)	1973	Gas	56	455	
Marbach, Germany (EVS Schwaben)	1973	Gas	55	260	
Type 2: Recuperation Siersdorf, Germany	1973	Gas (coke oven)	2×20	2×12	37·1 G

The seventh entry under '*Type 1: High efficiency*' should read as follows:

Vitry sur Seine, France (2 units)	1970	Gas and coal	42	280	42 G

Page 78. Left column, end of third paragraph. For 'but without fan debit' read 'but without f.d. (forced draught) fan debit'.

Page 79. Left column, fourth line. For 'and at a capital cost on F.P.C. rating' read 'and its capital cost on F.P.C. rating'.

Page 80. Fig. 76.6a. On horizontal line BC, for 'lb H_2O' read '1 lb H_2O'. The scale divisions were omitted on the left. A corrected figure is shown opposite.

Page 81. Fig. 76.6b. For 'ENTHALPY IN GAS—Btu' read 'ENTHALPY—Btu/lb H_2O'.

Page 81. Right column, third line. For 'in the region of 7000 h' read 'more nearly 4000 h'.

Page 82. Fig. 76.7b. Arrows to the curves were omitted. A corrected figure is shown opposite.

Page 84. Left column, fifth paragraph, fourth line. For 'superheater behind a convection surface' read 'superheater behind some convection surface'.

Page 84. Right column, fourth paragraph. For lines 8–12 read 'Drawbacks are, however, that the cost of the boiler increases fast with higher pressure; some fluids are expensive and some are toxic or objectionable in other ways; moreover, few complicated molecules are stable at temperatures as high as 600°F'.

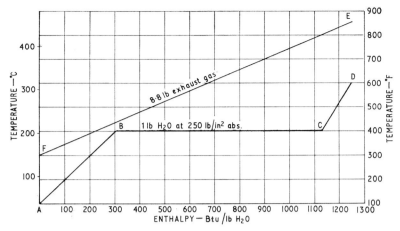

Fig. 76.6a

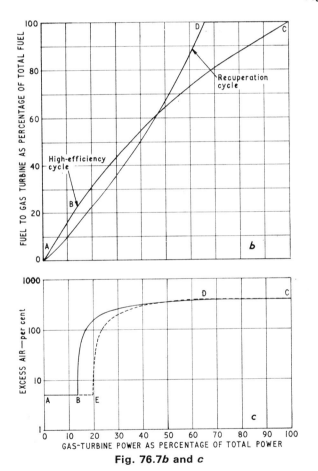

Fig. 76.7b and c

Page 85. Left column, third paragraph, fourth line. For '50 MW gas turbines' read '50–60 MW gas turbines'.

E. B. Johnson

My replies to the comments of F. Walker are as follows.

Boiler control systems. It is certainly true that there have been recurrent problems with some loops on boilers in the

first half of the 500 MW construction programme. However, these have often arisen from the performance of the plant rather than from control-system design. This is exemplified by the need for quantities of spray water much larger than the designed amounts on certain boilers, a fault affecting a sequence of possibly four stations all under construction before corrective measures could be taken. This single cause has led directly to a large number of the recurrent temperature-control problems. Additionally, the need to meet these requirements for spray water severely inhibits the correct operation of the feed control by requiring high-pressure differentials across the feed valves, often in excess of the range of the instrument provided. Many of these problems are well known and have been taken into account in current design work. However, the process of detection and rectification of plant and system faults on a new series of power plants taking four to six years to design and commission is necessarily a slow one.

Plant sensing devices. The problem of the application of limit switches to power plant is a difficult one. Many of the switches installed are of good design and are satisfactory when utilized in the correct applications, but prove very troublesome in adverse environments. In addition, many otherwise suitable devices have given trouble because of the method of installation and actuation. The best hope for a solution to the problem appears to be the use of approved limit switches and their installation in accordance with a code of practice prepared by control system, plant design and operation staff; this is a technique which we are now developing with the C.E.G.B.

Auto sequence control. I agree that it is important to provide facilities to enable the cause of sequence-control faults to be diagnosed quickly. However, I do not agree that these facilities need to be provided in the control room for use by the unit operator. Current practice in the C.E.G.B. is to provide no independent manual control from the control room for the individual items of plant covered by a sequence. Reliability of the sequence-control system is therefore all-important and it will not be possible to two-shift the plant if numerous random faults occur. If there is a single sequence fault, the operator is expected to initiate an alternative plant sequence if one is available, e.g. for a fan, mill or pump, whilst the maintenance staff rectify the defective sequence. To this end, full individual step-fault indication is provided at the sequence-control cubicles in the equipment room. In addition, the shutdown sequence is so arranged that its correct completion will provide a degree of pre-start check on the start-up sequence. A failure to complete will enable the shutdown hours to be used for rectification before hot starting.

The preceding remarks apply to digital on–off sequence steps such as are involved in a typical plant-item sequence, e.g. induced-draught fan start-up. In the automatic run-up and loading of the turbine a single parameter is continuously monitored and controlled, e.g. turbine speed or load, and we regard this as being more a modulating control function than a sequence. In this case, a number of safety measurements can initiate a turbine hold (vibration, eccentricity, temperature differentials), and these are indicated to the operator so that he can determine the next course of action, such as to trip the unit or to increase manually speed or load.

Are we tending to automate unnecessarily? There is not sufficient time during a routine hot start for the operator to start the auxiliaries relating to the second half of the draught plant and other major items without automation. In any case, if the traditional manual start-up of, say, an induced-draught fan is examined, it will be found that many of the actions involved are subjected to permissive interlocks which, unless cleared, will prevent start-up. In many cases these interlocks use the same plant sensing devices as are now used for sequence control. It can, therefore, be argued that, provided the automatic sequence control is not encumbered by unnecessary pre-start checks, its reliability should not be greatly below that of manual control.

The electric governor has been developed in order to give improved speed of response, and hence greater protection to the turbine against accidental overspeed. It is also designed to have a linear steam flow/frequency characteristic. In this respect, tests have shown that the existing mechanical governors vary widely in linearity and that some are very poor indeed. Variations in measured droop in the upper part (70–100 per cent) of the load range give typical changes of gain of 3:1.

The ability to choose alternative settings of governor proportional band is regarded as a potential system advantage of the electric governor rather than a major reason for its adoption.

Now here is my reply to K. Gorter.

(1) Prior to steam admission, fuel is normally held constant while steam temperature is rising and the steam mains are warming through. A check would be kept during this period on absolute metal temperatures, to prevent excessive rise.

After synchronizing, the loading rate is determined by thermal-stress considerations in the turbine h.p. and i.p. cylinders. Thermal stress is inferred from the usual measurements of cylinder inner and outer surface temperatures.

On the boiler we are more interested in the absolute tube temperatures because of 'creep-life' considerations.

(2) Feedwater controls are designed to be switched to 'automatic' as soon as a steady feedwater demand is established, usually just prior to synchronizing. Low load feed control is performed by a 20 per cent load valve capable of handling large pressure differentials. Changeover to the full-range system is arranged to take place automatically at about 20–25 per cent load.

Temperature controls are designed for a more limited range, but it has proved necessary to extend this to a lower load range on a number of boilers. Such an extension has required re-design of some superheater spray systems, replacement of some spray valves, and use of load-adaptive controllers.

Provided tight shut-off is obtained from the spray valves, there is no reason why the controls should not be on 'auto' prior to the need for spray water, but the control system must be designed to prevent 'integral wind-up'. Where reheat-steam temperature is controlled by gas by-passing, the damper controls must be switched to 'manual' below about 70 per cent load, otherwise reheat-steam temperature will tend to be maintained at its set-point at the expense of superheat. This problem can be overcome by automatic set-point scheduling but provision of this facility is still in the experimental stage.

(3) We prefer to run-up a machine to a maximum stress criterion as determined by the measurement of metal temperature gradient. It is not, therefore, possible to state a definite value of temperature difference between steam and metal, because this will vary with steam flow.

(4) A typical run-up time for a start after 6 h shutdown is 10 min.

(5) The whole development of large steam-turbine design in the United Kingdom has been based for many years on the requirement for quick starting and it is difficult at this stage to say what particular features are provided specifically to meet this requirement. The use of throttle governing with full steam admission at the h.p. turbine inlet is, however, regarded as important to facilitate uniform warming-up of the cylinder and particular attention is paid to the geometry of the casings to minimize thermal stresses during the process. The provision of adequate internal clearances and the arrangements for ensuring freedom of movement of the cylinder are equally important.

H. Czermak and K. Goebel

The following correction to Paper 89 should be noted.

Page 107. H. Czermak is Oberingenieur at Hohe Wand and employed by NEWAG.

C. Hijszeler and K. van Duinen

We certainly agree with K. Gorter that the design of the gas and air flow measurement system should be correct. Care should be given to evaluate the accuracy of air and gas flow measurements, particularly at low loads during the design stage. In our paper we wanted to stress the point that the adjustment of the fuel/air ratio does not normally limit the use of burners with a high capacity in relation to the total input.

With reference to his final remarks we would say that the natural gas delivered to the customers of the gas companies is normally dry. There may, however, occur abnormalities in the gas transport system and the gas distribution system of the customer that cause the introduction of liquids, dust or other materials in the firing installation. It is for this reason that the incorporation of a filter with liquid storage capacity in the gas supply system to the boiler is recommended by the gas companies.

In reply to W. J. Hoornweg, the problem of the evaluation of the accuracy of the measurement of a gas flow of a range as under discussion is that an absolute standard is not available. All discussions about accuracy are based on comparisons of measurements with unknown accuracies.

It is Gasunie's aim to improve the accuracy of flow measurements by the use of more accurate elements in the measurement system, e.g. an accurate density meter; the improvement of the installation technique; correction of the measurements by compensation for the calibration data.

The experience given in F. A. Allan's contribution is a valuable addition to the paper. Referring to the breakdown of the conversion costs we would observe that the breakdown given in the paper indicates an average. Deviations of the figures occur as F. A. Allan's example indicates.

In reply to O. Wiltshire, the safety rules in the Netherlands are made on the basis of the N.F.P.A. rules. The committee that made the rules has made adjustments in accordance with its own experience and opinion. Explanation of more details is beyond the scope of the paper.

The purpose of the bleed valve is to relieve the pressure between the block valves in case these are closed, thus reducing the possibility of gas leakage into the furnace. The flow indicator is installed to indicate whether there is a constant gas flow through the bleed pipe, indicating a leaking block valve.

Furthermore the bleed valve is used during the tightness test on the block valves.

The monitoring of each individual burner is regarded as the optimum solution to obtain safe operation and is done in the majority of the cases where multi-burner installations are in use. The ideal solution for monitoring a burner is to check the igniter for presence of flame before gas is allowed to enter the main burner; after lighting-up of the main burner the gas supply to the igniter is stopped and the flame on the main burner is checked.

In the Netherlands the purge of the furnace is based on the number of air changes. Up till now no necessity has been felt to deviate from this principle.

Serious furnace explosions have not occurred up till now with installations built according to the safety regulations.

One minor explosion occurred as a result of a steam leakage in a combustion air collector, which reduced the air quantity beyond the limit of stable operation of the burner.

Another explosion occurred during the starting-up of a boiler of a power station. In this case the installation had not been operated in compliance with the more recent safety rules owing to the early date of completion of this unit.

In the Netherlands the safety regulations to be followed are enforceable by law.

The following corrections to the paper should be noted.

Page (column)
114 (1, 2) K. van Duinen is an employee of Koninklijke Machinefabriek Stork N.V. Hengelo, Netherlands.
114 (2) The last line should read: 'maximum pressure of 67 bar, feeding regional grid systems operated at a maximum pressure of 40 bar. From these systems the supply'.
115 (2) Last line (and first line of p. 116): for 'of a very constant quality' read 'very constant'.
120 (2) The third line should read: '80–100°C by steam air heaters for the same reason. With sulphur-free natural gas the dewpoint of the flue gas is only 55–60°C as the result of the water vapour content. This'.
121 (1, 2) Fig. 97.7 should be replaced by the diagram overleaf.
121 (2) Seventh line from the bottom: for '100' read '110'.
122 (1) Fifth line: for 'value' read 'volume'.
122 (1) Section headed 'Coal fired in burners', third line from the bottom: for 'unnecessarily' read 'unacceptably'.

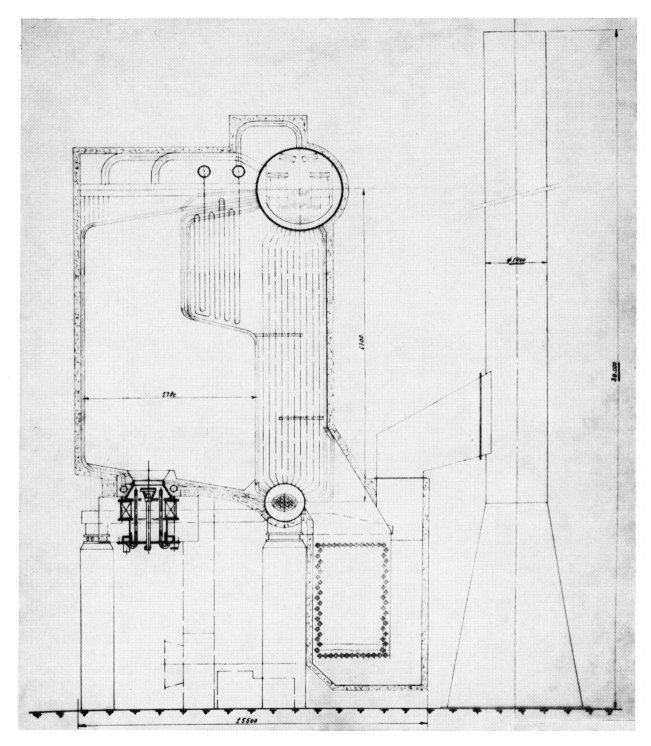

Fig. 97.7 (replacement). Stork T.D.G. type boiler

List of Delegates

ALBAN, F. P. — Kellogg International Corporation, London.

ALLAN, F. — P.G.E.M., Arnhem, The Netherlands.

AWMACK, C. W. — Ministry of Defence (Navy), Bath, Somerset.

BARR, D. C. — Electro-Watt Engineering Services, Ltd, Zurich, Switzerland.

BELL, A. — Foster Wheeler John Brown Boilers Ltd, London.

BERTRUMS, F. S. — N.V. Interutil, Vlaardingen, The Netherlands.

BLAAUW, R. — Provinciaal Electriciteitsbedrijf van Noord-Holland, Bloemendaal, The Netherlands.

BONSEL, W. G. — Reactor Centrum Nederland, Petten, The Netherlands.

BRETT, R. E. — Weir Pumps Ltd, London.

BROEZE, J. J. — Technical University of Delft, The Netherlands.

BURGER, C. P. — Koninklijke Machinefabriek Stork N.V., Hengelo, The Netherlands.

CABANES, J. P. — La Maquinista Terrestre y Maritime S.A., Barcelona, Spain.

CARTER, N. W. T. — Clarke Chapman–John Thompson, Durham.

CHRISTOPHERSEN, M. C. — Isefjord Power Company, Copenhagen.

CORNWELL, K. J. — Heriot-Watt University, Edinburgh.

COSAR, P. — Babcock-Atlantique, Paris.

CURRIE, J. W. — South of Scotland Electricity Board, Glasgow.

CZERMAK, H. — Niederösterreichische Elektrizitatswerke A.G., Enzerdorf, Austria.

DEEN, P. J. — Shell Internationale, 's Gravenhage, The Netherlands.

DENNETT, R. L. — Foster Wheeler John Brown Boilers Ltd, London.

DE HES, M. — N.V. Neratoom, 's Gravenhage, The Netherlands.

DE JONG, J. A. — Technical University of Delft, The Netherlands.

DIJKHOFF, P. — Gemeente-Energiebedrijf, Amsterdam.

DOLEZAL, R. — Sulzer Brothers Ltd, Winterthur, Switzerland.

DYKES, I. M. — Weir Pumps Ltd, Glasgow.

EWBANK, M. H. — Ewbank and Partners Ltd, Brighton, Sussex.

FULLER, J. A. — Hopkinsons Ltd, Huddersfield, Yorkshire.

GOLDSACK, J. S. — Babcock and Wilcox (Operations) Ltd, London.

GOLDSTERN, W. — Verein Deutscher Ingenieure, Essen, Germany.

GORTER, K. — P.G.E.M., Arnhem, The Netherlands.

GOTT, H. H. — Associated Nuclear Services, London.

GOEBEL, K. — Kraftwerk Union A.G. Erlangen, West Germany.

GRAVENBERCH, W. R. — Koninklijke Machinefabriek Stork N.V., Hengelo, The Netherlands.

GROLLE, J. — The Rotterdam Dockyard Company, Rotterdam, The Netherlands.

GREEN, R. A. W. — Central Electricity Generating Board, London.

HAERSCHNITZ, H. A. O. — Koninklijke Machinefabriek Stork N.V., Hengelo, The Netherlands.

HAFTKE, J. J. — Babcock & Wilcox (Operations) Ltd, London.

HARKNESS, W. A. — Merz and McLellan, Newcastle upon Tyne.

HAYDEN, R. L. J. — Foster Wheeler John Brown Boilers Ltd, London.

HAYES, C. W. — W. H. Allen Sons & Co Ltd, Bedford.

HILLIS, R. W. — Foster Wheeler John Brown Boilers Ltd, London.

HIRST, A. W. C. — Rugby, Warwickshire.

HOEKSTRA, P. E. — N.V. Motorenfabriek Thomassen, De Steeg, The Netherlands.

HOLLICK, W. — Unilever Ltd, London.

HOLLINS, R. T. — Bailey Meters & Controls Ltd, Croydon, Surrey.

HONEYWELL, C. R. J. — British Leyland Motor Corporation, Abingdon, Berks.

HOORNWEG, W. J. — P.G.E.M., Arnhem, The Netherlands.

HUMPHREYS, L. J. — Electricity Supply Board, Dublin 2.

HYSZELER, C. — N.V. Nederlandse Gasunie, Groningen The Netherlands.

INCHES, R. M. — Admiralty Marine Engineering Establishment, Gosport, Hants.

JOHNSON, E. B. — Central Electricity Generating Board, London.

KANTNER, A. — Vereinigung der Grosskesselbetreiber E.V., Essen, West Germany.

KASSEBOHM, B. — Stadtwerke Dusseldorf, Dusseldorf, West Germany.

KAT, D. — Rontgen Technische Dienst N.V., Rotterdam, The Netherlands.

KERKHOVEN, H. — E.M. Electrostoom N.V., Rotterdam, The Netherlands.

KOK, H. — E.M. Electrostoom N.V., Rotterdam, The Netherlands.

KONINGS, J. A. — N.V. Motorenfabriek Thomassen, De Steeg, The Netherlands.

KOOPMAN, J. W. — Provinciaal Electriciteitsbedrijf van Noord-Holland, Bloemendaal, The Netherlands.

LAMBERT, H. J. — De Rotterdamsche Droogdok Maatschappij N.V. Rotterdam, The Netherlands.

LAURSEN, N. — Isefjord Power, Copenhagen.

LAWRIE, P. — P. & O. Steam Navigation Co. Ltd, London.

LINDO, A. E. — P.G.E.M., Arnhem, The Netherlands.

LISOWSKI, J. — Peabody Ltd, London.

MARSHALL, K. — Balfour Beatty Co. Ltd, Sidcup.

McOWAT, P. — South of Scotland Electricity Board, Glasgow.

MEJNERTSEN, S. A. — Isefjord Power, Copenhagen.

MERCER, S. — Solihull, Warwickshire.

MEYLER, J. J. — Electricity Supply Board, Dublin 2.

MIJNARENDS, H. — Elektriciteitsbedrijf, Delft, The Netherlands.

MITCHELL, J. M. — C. A. Parsons & Co. Ltd, Newcastle upon Tyne.

MITCHELL, R. W. S. — Technische Hogeschool, Delft, The Netherlands.

MORELLE, J. — Worthington Ned.N.V., 's Gravenhage, The Netherlands.

MORRIS, H. — South of Scotland Electricity Board, Glasgow.

MUYSKEN, M. — Reactor Centrum Nederland, 's Gravenhage, The Netherlands.

NAJMAN, B. B. — Sulzer Brothers Ltd, Winterthur, Switzerland.

NICHOLS, R. W. — Department of Trade and Industry, London.

NONHEBEL, G. — Romsey, Hants.

PACZUSKI, J. S. — Babcock & Wilcox Ltd, London.

PARSONS, N. C. — C. A. Parsons & Co. Ltd, Newcastle upon Tyne.

POLMAN, F. E. — E.M. Electrostoom N.V., Rotterdam, The Netherlands.

POTTERILL, R. H. — Ewbank & Partners Ltd, Brighton, Sussex.

POWELL, B. — Central Electricity Generating Board, Birmingham.

RATCLIFFE, F. — Central Electricity Generating Board, Hastings, Sussex.

ROBEY, L. A. — Bowater Paper Corporation, London.

ROOK, R. A. — De Rotterdamsche Droogdok Maatschappij N.V., Rotterdam, The Netherlands.

ROOZENDAAL, G. — P.G.E.M., Arnhem, The Netherlands.

SALMON, A. — Clarke Chapman–John Thompson Ltd, Wolverhampton, Staffs.

SCHMIDT, F. H. — Lummus Nederland N.V., 's Gravenhage, The Netherlands.

SCHÜRMANN, L. J. H. — G.E.B. Dordrecht, The Netherlands.

SCOTT, R. W. — British Nuclear Forum, London.

SHEPHERD, J. — Central Electricity Generating Board (Northern Project Group), Cheshire.

SHERRY, A. — Central Electricity Generating Board, Ipswich, Suffolk.

SHILLING, W. F. — Gulf General Atomic Europe, Zurich, Switzerland.

SPRAGUE, D. W. — United Kingdom Atomic Energy Authority, Thurso, Caithness.

STIBBE, P. W. — N.V. Koninklijke Maatschappij De Schelde, Vlissingen, The Netherlands.

STIGTER, D. — N.V. Electriciteits-Maatschappij, Zwolle, The Netherlands.

STINSON, W. H. — Westinghouse Electric Corporation, Philadelphia, Pa.

SUYKER, Th. W. — M/s A. Kiewit, Rotterdam, The Netherlands.

TOMLIN, R. J. — BP Chemicals International Ltd, Port Talbot, Glam.

TUININGA, P. — 's Gravenhage, The Netherlands.

UITTENBOSAART, J. — Delta Engineering N.V., Schiedam, The Netherlands.

UPTON, D. E. — Babcock & Wilcox (Operations) Ltd, London.

VAN DEN BERG, J. M. — Provinciaal Electriciteitsbedrijf van Noord-Holland, Bloemendaal, The Netherlands.

VAN DEN HOOGEN, B. — Nederjandsche Electrolasch Maatschappij N.V., Leiden, The Netherlands.

VAN DE STADT, H. — Gemeenhelijk Enerjiebedrijf, 's Gravenhage, The Netherlands.

VAN DER SCHRIER, W. J. — Geveke & Groenpol, Amsterdam.

VAN DUINEN, K. — Koninklijke Machienfabriek Stork N.V., Hengelo, The Netherlands.

VAN ERPERS ROYAARDS, R. — N.V.K.E.M.A., Arnhem, The Netherlands.

VAN HOEK, G. A. L. — N.V. Sep, Arnhem, The Netherlands.

VAN KUIJK, R. M. — N.V. K.E.M.A., Arnhem, The Netherlands.

VAN LOOSEN, M. — N.V. Koninklijke Maatschappij De Schelde, Vlissingen, The Netherlands.

VAN MELICK, F. A. W. H. — N.V. Provinciale Noordbrabantsche 's Hertogenbosch, Electriciteits-Maatschappij, The Netherlands.

VAN REYEN, G. — Commission of the European Communities, Brussels.

VAN STAA, H. — Technische Hogeschool, Delft, The Netherlands.

VAN VEEN, M. C. — Rotterdam Dockyard Company, Rotterdam, The Netherlands.

WALKER, F. — South of Scotland Electricity Board, Glasgow.

WHILLOCK, R. G. — Preece, Cardew & Rider, Brighton, Sussex.

WIDMER, M. — Cie Electro-Mécanique, Le Bourget, France.

WILKINSON, T. J. — Central Electricity Generating Board (Generation Development Construction Division), London.

WILTSHIRE, O. — Kennedy & Donkin, Manchester.

WOOD, B. — Merz and McLellan, Esher, Surrey.

WYLHUIZEN, H. — K.E.M.A., Arnhem, The Netherlands.

YOUNGS, J. A. — Brown Boveri International, Baden, Switzerland.

Index to Authors and Participants

The names of authors and the numbers of pages on which papers begin are printed in bold type.

Subject Index

Titles of papers are in capital letters.

MADE AND PRINTED IN GREAT BRITAIN BY WILLIAM CLOWES & SONS, LIMITED, LONDON, BECCLES AND COLCHESTER